U0114023

NONLINEAR PHYSICAL SCIENCE
非线性物理科学

NONLINEAR PHYSICAL SCIENCE

Nonlinear Physical Science focuses on recent advances of fundamental theories and principles, analytical and symbolic approaches, as well as computational techniques in nonlinear physical science and nonlinear mathematics with engineering applications.

Topics of interest in *Nonlinear Physical Science* include but are not limited to:

- New findings and discoveries in nonlinear physics and mathematics
- Nonlinearity, complexity and mathematical structures in nonlinear physics
- Nonlinear phenomena and observations in nature and engineering
- Computational methods and theories in complex systems
- Lie group analysis, new theories and principles in mathematical modeling
- Stability, bifurcation, chaos and fractals in physical science and engineering
- Discontinuity, synchronization and natural complexity in physical sciences
- Nonlinear chemical and biological physics

SERIES EDITORS

Albert C. J. Luo
Department of Mechanical and Mechatronics Engineering
Southern Illinois University Edwardsville
IL 62026-1805 USA
Email: aluo@siue.edu

Dimitri Volchenkov
Department of Mathematics and Statistics
Texas Tech University
1108 Memorial Circle, Lubbock
TX 79409 USA
Email: dr.volchenkov@gmail.com

INTERNATIONAL ADVISORY BOARD

J. A. Tenreiro Machado · Dimitri Volchenkov *Eds*.

Mathematical Topics on Modelling Complex Systems

In Memory of Professor Valentin Afraimovich

复杂系统建模的数学问题

纪念Valentin Afraimovich教授

Higher
Education
Press

Editors

J. A. Tenreiro Machado
ISEP-Institute of Engineering
Polytechnic Institute of Porto
Porto, Portugal

Dimitri Volchenkov
Department of Mathematics and Statistics
Texas Tech University
Lubbock, TX, USA

图书在版编目（CIP）数据

复杂系统建模的数学问题 = Mathematical Topics on Modelling Complex Systems: In Memory of Professor Valentin Afraimovich : 英文 /（葡）滕雷罗·马查多 (J.A.Tenreiro Machado)，（美）迪米特里·沃尔琴科夫 (Dimitri Volchenkov) 主编 . -- 北京：高等教育出版社 , 2023.8

（非线性物理科学）
ISBN 978-7-04-060822-9

Ⅰ . ①复… Ⅱ . ①滕… ②迪… Ⅲ . ①系统建模—英文 Ⅳ . ① N945.12

中国国家版本馆 CIP 数据核字 (2023) 第 125481 号

策划编辑　吴晓丽　　责任编辑　吴晓丽　　封面设计　杨立新
责任印制　韩　刚

出版发行	高等教育出版社	网　　址	http://www.hep.edu.cn	
社　　址	北京市西城区德外大街 4 号		http://www.hep.com.cn	
邮政编码	100120	网上订购	http://www.hepmall.com.cn	
印　　刷	涿州市星河印刷有限公司		http://www.hepmall.com	
开　　本	787 mm×1092 mm　1/16		http://www.hepmall.cn	
印　　张	12.5			
字　　数	360千字	版　　次	2023 年 8 月第 1 版	
购书热线	010-58581118	印　　次	2023 年 8 月第 1 次印刷	
咨询电话	400-810-0598	定　　价	119.00 元	

本书如有缺页、倒页、脱页等质量问题，请到所购图书销售部门联系调换
版权所有　侵权必究
物　料　号　60822-00

Preface

The present edited volume explores recent developments in theoretical research on mathematical topics on modelling complex systems. The volume is dedicated to the memory of our colleague Valentin Afraimovich (1945–2018), a visionary scientist, respected colleague, generous mentor, and loyal friend. Professor Afraimovich was a Soviet, Russian, and Mexican mathematician known for his works in dynamical systems theory, qualitative theory of ordinary differential equations, bifurcation theory, concept of attractor, strange attractors, space-time chaos, mathematical models of nonequilibrium media and biological systems, traveling waves in lattices, complexity of orbits and dimension-like characteristics in dynamical systems.

The collection of works in this edited volume opens with the calligraphy work in memory of Prof. Afraimovich performed by Prof. Albert Luo. The first part of the volume is devoted to the mathematical tools for the design and analysis in engineering and social science study cases. We discuss the periodic evolutions in nonlinear chemical processes, vibro-compact systems and their behavior, different types of metal-semiconductor self-assembled samples, made of silver nanowires and zinc oxide nanorods, and evolution of systems with power-law memory. The second part of the book is devoted to mathematical description and modelling of the critical events, climate change, and robust emergency scales. In three chapters, we consider a climate-economy model with endogenous carbon intensity and the behavior of Tehran Stock Exchange market under international sanctions. The third part of the book is devoted to fractional dynamic and fractional control problems. We discuss the novel operational matrix technique for variable-order fractional optimal control problems, the nonlinear variable-order time fractional convection-diffusion equation with generalized polynomials, and solvability and inverse problems in differential and integro-differential equations. The book facilitates a better understanding of the mechanisms and phenomena in nonlinear dynamics and develops the corresponding mathematical theory to apply nonlinear design to practical engineering.

Valentin Afraimovich was a generous, gregarious, energetic presence at the very heart of nonlinear dynamics and complexity science communities, all of which were transformed by his presence. We hope that the scientific community will benefit from this edited volume.

Lubbock, TX, USA   Dimitri Volchenkov
Porto, Portugal, European Union   J. A. Tenreiro Machado

In Memory of Professor Valentin Afraimovich (1945–2018)

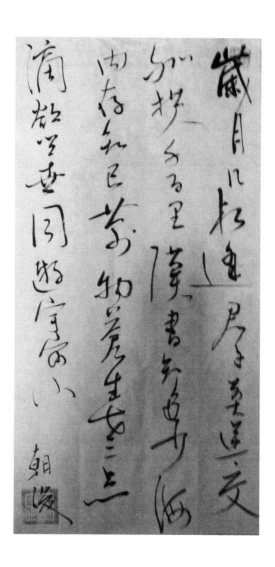

Even though time goes years by years, how many people can meet each other?

How come we have a permanent, gracious friendship for many years? This is because you and I have the same dreams.

Indeed, you and I have a long journey in science which liked that we ride on the flying horses to fight together.

After looked at the fantastic stories in the past, how many you and I discovered in science can be written in the history book?

In the world, you knew me, and I knew you. How many people can understand each other in this world!

Please do not be sad, my friend, every living thing will become old when time passes.

I believe, our little contributions in science will last forever to this world. For the human-being progress, we did what we did.

Because of you and me in this world, our contributions made the universe be less mysterious. Bye-bye, my friend!

Albert C. J. Luo

Contents

Inverse Problems for Some Systems of Parabolic Equations 163
C. Connell McCluskey and Vitali Vougalter

**Solvability in the Sense of Sequences for Some Non-Fredholm
Operators with Drift** 171
Vitaly Volpert and Vitali Vougalter

Applications of the Fundamentals of Bézier Curves

Mayra Angélica Bárcenas-Castro, Ramón Díaz de León-Zapata,
Saúl Almazán-Cuéllar, Efrén Flores-García, Gustavo Vera-Reveles,
and José Vulfrano González-Fernández

Abstract This study presents applications of Bézier curves as design and analysis tools in two engineering study cases. In the first case, different types of metal–semiconductor self-assembled samples, made of silver nanowires and zinc oxide nanorods, were electrically and thermally characterized (Resistance & Temperature) according to the main merit figure for bolometric detectors. The Thermal Coefficient of Resistance (TCR) was used in order to assess their performance as bolometric materials. The representation of the graphical behavior was obtained using spline interpolation, applied for three samples of different concentration of Ag/ZnO, smoothing the obtained curve, results showed that higher ZnO concentrations will result in films with higher resistance and higher TCR, having the possibility to tune important parameters such as the TCR and film resistance by changing the ZnO concentration of the sample. In the second case, Bézier curves are controlled by means of genetic algorithms to obtain the maximum concentration or emission of electromagnetic radiation in a specific frequency range of the nanoantenna under study. In this way, Bézier curves have been presented as an optional tool for geometric manipulation assisted by computer in specialized engineering applications that could be useful in similar works.

Keywords Bézier curves · Spline curves · Geometric variations

1 Introduction

Bézier curves turn out to be a tool in the treatment of polynomial surfaces; some relevant applications can be found in computer-aided geometric design [1], trajectory generation [2] and reconstruction of models [3]. The Bézier curve is defined between two external points (starting point and end point) using one or more points called control points that deform the curve when trying to join all the points involved (starting point, end point and control points). If the Bézier curve uses a single control

M. A. Bárcenas-Castro (✉) · R. Díaz de León-Zapata · S. Almazán-Cuéllar · E. Flores-García ·
G. Vera-Reveles · J. V. González-Fernández
Tecnológico Nacional de México/IT de San Luis Potosí, Av. Tecnológico s/n, 78437 Soledad de
Graciano Sánchez, SLP, México

1

Fig. 1 Second degree
Bézier curve

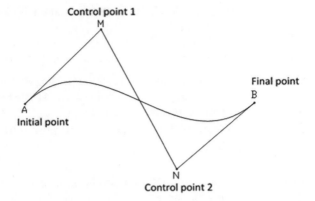

point, it is called a first-degree Bézier curve, while if it uses two control points, it is called the second-degree Bézier curve and so on, according to the number of Bézier curves, as shown in Fig. 1.

The union of several of these curves can form an open or closed geometry, with as many curves as necessary to achieve the desired resolution or smoothing effect (quality) of the final geometry. In this sense, flexible polynomial curves for conic sections are called rational Bézier curves.

The Bézier curve of n-degree is defined as:

$$b(t) = \sum_{i=0}^{n} b_i B_i^n(t), \quad (0 \le t \le 1) \ with \ B_i^n(t) = \binom{n}{i} t^i (1-t)^{n-i} \quad (1)$$

where $b_i \in \mathbb{R}^2$ in a set of $n+1$ control points are required and $B_i^n(t)$ is the Bernstein polynomial of n-degree [4].

Iso Schoenberg, considered the father of Splines, wrote: 'Polynomials are wonderful even after they are cut into pieces, but the cutting must be done with care. One way of doing the cutting leads to the so-called spline functions', in 1964 [5], Splines is a generalized curve of Bézier curves [6]. Spline curves are an interpolation method that results in a smooth surface that passes exactly through the desired points [7], which give similar results, requiring only the use of low-degree polynomials, thus avoiding undesirable oscillations in most of the applications, found when interpolating by polynomials of high degree [8].

A cubic Spline function is defined with the following expression:

$$S_i(x) = \alpha_i + \beta_i(x - x_i) + \chi_i(x - x_i)^2 + \delta_i(x - x_i)^3 \quad (2)$$

where $S_i(x)$ is the cubic function for a segment, x is the first point of the segment, x_i is the function evaluation point ($i = 1, 2, \ldots, n$) and $\alpha_i, \beta_i, \chi_i, \delta_i$ are the coefficients of the function [9].

In this context, Bézier curves, unlike other commonly used similar functions such as interpolations, are an excellent tool. For example, in the industrial sector it could be applied for computational design of parts or minimizing interpolations and its constructing graphs for interpretation; so, a high computational cost is not required [10].

2 Methodological Isomorphism; Bézier Curves and Their Applications

Analyzing topological structures allows to generate information about known elements. One way to establish a defined concept in a different study area is by means of isomorphisms [11]. In this sense, the greatest number of properties on these geometries under study. In this work, two applications for engineering are presented.

Application 1: Spline curve applied to thermoelectric characterization in Ag/ZnO bolometers. One of the most important thermal detectors is bolometer, which is a temperature-sensitive electrical resistor, as a consequence, a significant change in the electrical resistance occurs when the detector is heated by incident infrared radiation measured by an external electrical circuit [12, 13]. In practice, the bolometer is made of a thin metal or semiconductor film, with an absorption film deposited on it, which is usually above a suspended structure that provides thermal isolation. Bolometric detectors are primarily used for the detection of far infrared and terahertz radiation, is of great relevance for a growing list of imaging, biomedical and industrial applications. Several examples are early detection of biological abnormalities in humans as well as in animals and living tissue samples, night vision sensors, early warning fire detection systems, and search and rescue equipment [14].

The electrical resistance of the film is controlled, while the IR radiation is absorbed and its temperature increases. If the film is metallic, the resistance increases with an increase in temperature and TCR is assigned; if the film is a semiconductor, the resistance decreases as the temperature rises, it is said to have a negative temperature coefficient of resistance [15]. When the radiation is removed, the temperature of the bolometer returns to its initial value, which is determined by the ambient surroundings in which it is immersed.

Generally, bolometers have smaller specific sensibility than optoelectronic, photoconductive or photovoltaic devices, however these last typically require cooling to operate efficiently.

The TCR is calculated by using the following formula:

$$\alpha = \frac{1}{R}\frac{dR}{dT} \tag{3}$$

where R represents the resistance and T represents the temperature.

Previous investigations suggest that ZnO indeed has higher TCR values than most common materials used in bolometric applications, like vanadium oxide (VOx) and amorphous silicon (a-Si) [16].

In reported works, ZnO films have been deposited using pulsed laser deposition and sputtering. In these samples the observed TCR values ranging from −3.4 to − 13% K^{-1}, however, these values were measured at temperatures much lower than room temperature [17]. Other works have shown that an increase in TCR can be obtained by using nanostructured materials [18–20]. The detailed sample preparation and device fabrication are described elsewhere [21–24].

TCR was calculated by three points averaging method [25] with the Eq. (3), given the set of points resulting from TCR, called control points. It is possible to observe in each of the three graphs of Fig. 2 a smooth surface that crosses exactly through the given points, thus it is worth mentioning that the spline interpolation used in this work, allows to graphically observe the absolute maximum and absolute minimum points respecting TCR, indicating the maximum and minimum change in the temperature-dependent electrical resistor of each sample, which translates into having a better bolometric performance, as Table 1 shown. The first graph, observed from left to right is from a sample consisting of pure silver nanowires, the TCR was found 1.04% K^{-1}, Standard Deviation (SD) = 0.23% K^{-1}, at 294 K, the positive TCR value is indicative of a conductor bolometer. The second graph is from a sample that was prepared from Ag/ZnO nanostructures, and showed a TCR of −9.69% K^{-1}, SD = 3.38% K^{-1}, at 302.7 K. Finally, the third graphic is from a sample that was prepared by using ZnO nanoparticles as seeds increasing the ZnO concentration, which is higher than

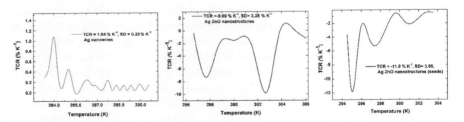

Fig. 2 Temperature-dependent TCR, Spline curve smoothing was used for the different types of characterized samples, where the Ag/ZnO nanostructures (seeds) possess TCR value = −11.8% K^{-1}, higher (compared to characterizations of this same material) reported at room temperature

Table 1 Shows the maximum TCR and SD of the Ag + ZnO respectively, as well as the temperature level at which the maximum TCR has been obtained

Sample	Maximum TCR (% K^{-1})	SD(% K^{-1})	Temperature at which TCR is maximum (K)
Ag nanowires	1.04	0.23	294.0
Ag/ZnO nanostructures	− 9.69	3.38	302.7
Ag/ZnO nanostructures (seeds)	− 11.80	3.09	295.0

the highest reported value for TCR $= -11.8\%$ K^{-1} in an Ag/ZnO film at room temperature, SD $= 3.09\%$ K^{-1} at 295 K. The negative TCR value of the last two samples are indicative of a semiconductor bolometer.

It is likely that by using ZnO nanoparticles as seeds, the increase in ZnO concentration, as well as the increase in the semiconductor material result in films with higher resistance and higher TCR. The SD obtained in the characterized samples, is close to a quarter of the maximum value for the conductive material of Ag nanowires and close to a third of the maximum values for semiconductor materials of Ag/ZnO nanostructures.

The advantage of using spline interpolation compared to other interpolation methods is that it requires the use of low-degree polynomials, at the same time avoiding undesirable oscillations that appear when interpolating by high-degree polynomials, also has a smaller error than some other interpolating polynomials such as Lagrange polynomial and Newton polynomial.

The Ag/ZnO concentration ratio influences the electrical and thermal characteristics of the deposited samples. Materials with higher TCR values can provide lower noise equivalent temperature difference (NETD) performance and increase the efficiency of the bolometers [26–28].

Application 2: Bézier curves controlled by genetic algorithms in optimized continuous geometries. The geometry or figure for many fields of engineering plays a fundamental role in the devices' performance as explained in [29, 30], just to mention a couple of cases. In the particular case of the nanotechnology field (but not limited to it), the behavior of electromagnetic radiation with matter depends significantly on the geometry [31] and it has been shown that the shape of these structures, to be optimal in some of the parameters of interest, do not necessarily follow the geometries that are often used in macroscopic antennas, such as those used in telecommunications [32].

On the other hand, the search for geometries that optimize parameters of interest have led to the application of optimization algorithms such as evolutionary algorithms.

There are other algorithms that could offer a mechanism for finding geometries or parameters considered as optimal, such as machine learning, deep learning among others.

Broadly speaking, learning-based algorithms can be classified into those that require a master on how to identify whether they are on the right way to find the optimal solution, or sophisticated algorithms that demand more computational cost. In the first case, usually a person indicates the more useful cases to the algorithm. This kind of interventions induce a low process rate of huge data volumes and the well know 'human error'. In the second case, learning without a master, the information of the possible optimal solutions must be taken from other sources, for example applying data mining. It also requires an additional coding effort that is not required when evolutionary algorithms are used.

Finally, in the case of artificial intelligence, where the techniques described above are usually applied, in addition to the evolutionary algorithms or maybe more techniques, would result in excessive computational cost for obvious reasons and would not necessarily provide significantly outstanding results, either in computational time or in optimization accuracy. It is important to say that the so-called analytical techniques are not applicable to these kinds of problems, since the final solution is not known and therefore there is not a mathematical model formalism to describe the dynamics of the systems under research.

Evolutionary algorithms do not require sophisticated learning processes or complex sub-algorithms of pre- or post-processing. These kinds of algorithms reveal their practical utility since they take the evolutionary process of the nature as a model, making the best adapted species survive to their environment, with the additional advantage that a computation system allows to perform in a few seconds, which can take nature millions of years. Even so, the selected algorithms for the study of nanoantennas have been the so-called genetic algorithms, since they are the ones that really take the model of nature.

Genetic algorithms as part of evolutionary ones, such as mono and multi-objective [32, 33] with a novel technique is exposed in which the geometric modeling by evolutionary algorithms ceases to be the placement or absence of material (individuals in the population) that generated discontinuous patterns, by a smooth and continuous geometry.

In the particular case of nanoscience and nanotechnology, because of the dimensions of this field of study, this technique allows an ease manufacture, as well as the continuity of material, which avoids possible interactions or parasitic resonances (plasmonics) in discontinuous cases, among others.

In the case of micro and submicrometric geometries, the individual is considered as one of the points of which a Bézier curve consists and a complete geometry will be made up of several of these curves. Each individual consists of a genetic load called genome, which is contained in its chromosomes. For the case of this proposal, the chromosomes are considered geometric positions in two or three dimensions depending on the application, and will be named as x, y and z.

In nanoantennas built by self-assembly techniques [21, 34] the thin surface is used and it is not relevant to consider the thickness within the optimization variables, however for the case of plasmonics [35, 36], as well as for some cases of metasurfaces design [37], the height (z) of the geometry is a parameter that must be considered.

Assuming the simplest evolutionary model of the genetic algorithm, the so-called "Holland genetic algorithm" [38] (although others may be applied depending on the particular application), a population of at least ten times the number of original individuals forming a sufficiently flexible pattern (a number of Bézier curves such that an arbitrary geometry can be molded) is initiated. For example, for the construction of a nano-type dipole, 16 Bézier curves can be enough, each one of third degree, that is with three control points in addition to their respective starting and ending points. This amount of Bézier curves (arbitrary at the moment) is under analysis by this same research group to quantitatively determine the minimum value of curves needed and is part of the future work to this contribution.

Ten times the amount of population of original individuals is needed to build an initial geometry that ensures genetic variation and avoid unhealthy individuals during the crossing that do not contribute to the convergence of the so called "fitness function".

The fact that for any antenna (including nanoantennas) the objective is that they present the maximum concentration or emission of electromagnetic radiation in a specific frequency range, one could try to select as a fitness function the maximum emission or reception according to the radiation source. However, these structures at the nanometric level shows amplified resonances that are not necessarily the maximum if their simulated or experimental value is unknown, so the health function will be changed by which represents the minimum losses in the frequency range of interest and, depending on the dimensions of the nanoantenna, thus leaving the Eq. (4) as the fitness function:

$$f_{fitness} = min\left(\frac{1}{2}Re(\sigma \mathbf{E} \cdot \mathbf{E}_{tot})\right) \qquad (4)$$

where

$$\mathbf{E}_{tot} = \mu_0 \int \left(\mathbf{J} + \varepsilon_0 \frac{\partial \mathbf{E}}{\partial t}\right)da \qquad (5)$$

J_{tot} is the total electric current density over the whole geometry obtained by the genetic algorithm,

$$J_{tot} = \sigma \mathbf{E} \qquad (6)$$

σ is the electrical conductivity and $\mathbf{E}_{tot}$ is the electric field over the whole geometry shown in Eq. (4).

As in any ecosystem, there is the possibility that individuals suffer mutations. These alterations to chromosomes are introduced into the computational algorithms to prevent the algorithm from being cycled, that it does not achieve convergence or that it is at a different point in the function to its actual maximum or minimum, what is usually known as false convergence.

The assigned value in percentage of mutation is usually very different between biological species. For example, higher animals such as humans, statistically have an extremely low mutation rate, while certain bacteria and viruses have a high one. In practice, this value is usually empirically adjusted after executing the algorithm a few times in order to establish the one that best corresponds, although a value present in many practical cases has been 0.1%, which perfectly fits the expected results for the tests carried out in this work.

There are other biological analogies for genetic algorithms, however it is not the objective here to justify or explain them all, but Table 2 presents a summary of the variables and their respective values.

Figure 3 shows the results of the algorithm execution in which the geometries

Table 2 General values for the experimental genetic algorithm

Parameter	Value
Population size	200
Elite count	0.05*PopulationSize
Crossover fraction	0.8
Stop criteria	100*number of variables or 0.01% of loses (first thing happens)
Number of variables	22 (Bézier control points and start and end curve points)

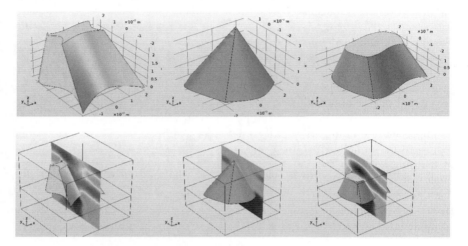

Fig. 3 Above and from left to right, initial geometry, some intermediate and final. Down, their respective responses to the presence of an electromagnetic field of three-dimensional nanostructures (plasmons)

obtained from the first iteration to the last one for three-dimensional geometries (plasmons) are appreciated since they are the most general, however it is easily applicable to two-dimensional cases (nanoantennas).

For the case of surface plasmons, unless it is another design situation, it is intended that most of the electromagnetic field is concentrated to the plane of incidence on the normal surface, which in this case is normal to the $x - y$ plane (from there, the name of plasmon surface resonance) and as can be seen in Fig. 3, the first iteration, which generated a completely random geometry, is able to significantly concentrate the electromagnetic field, but not to the plane of incidence on its normal surface. As the geometry evolved, an almost conical geometry was generated, but with a minimum concentration of electromagnetic field, behaving as a reflective surface.

Finally, in the last iteration, the maximum concentration at the normal surface of the plasmon is observed (intense red at the center) besides not having the same height of the initial or intermediate geometries, since this parameter also plays an important role in the multiobjective optimization.

3 Conclusions

The Bézier curves are a tool that can be applied to several scientific areas; in this work two cases of application were presented in engineering. The analysis and use of these will depend on the study phenomenon. In the first case, it was possible to visualize in more detail the graphic behavior applied by the Spline curves, to the bolometric performance given by the TCR, the softness of the curves is shown, as well as the absolute maximum and minimum values; in addition, by changing the concentration of ZnO in the sample, it is possible to better match thin bolometric samples to read-out integrated circuits. In the second case, the genetic algorithm generates certain geometries, so that the last iteration could be noticed the maximum concentration on the normal surface of the plasmon, besides observing that it does not have the same height of the initial or intermediate geometries, because this parameter also plays an important role in multiobjective optimization. In this sense, the health function represents the minimum losses in the range of frequencies of interest and in function of the dimensions of the nanoantenna under study. On the other hand, there are other mechanisms such as machine learning, deep learning or artificial intelligence that offer options to optimize the geometric trajectory of information. However, the Bézier curves applied to genetic algorithms turn out to be a clear example of low computational cost to be applied in a wide market of similar works.

References

1. Gómez Collado MDC, Puchalt Lacal J, Sarrió Puig J, Trujillo Guillen M (2013) Diseñar una obra en arquitectura desde un punto de vista matemático. Pensamiento Matemático 3(1):49–58
2. Sanchez CM, Sanchez JRG, Cervantes CYS, Ortigoza RS, Guzman VMH, Juarez JNA et al (2016) Trajectory generation for wheeled mobile robots via Bézier polynomials. IEEE Lat Am Trans 14(11):4482–4490
3. Antón DT, Jover FJE, Oviedo TP, Montserrat JMM, España M de P de V (2011) La restauración virtual de piezas arqueológicas a partir de datos procedentes de escáner 3D: reconstrucción volumétrica de una jarrita islámica del Museo Arqueológico Municipal de la Plana Baixa-Burriana (Castellón). III Internacional de Arqueología, Informática Gráfica, Patrimonio e Innovación
4. Prautzsch H, Boehm W, Paluszny M (2013) Bézier and B-spline techniques. Springer, Berlin Heidelberg
5. Schoenberg IJ (1964) On Trigonometric spline interpolation. J Math Mech 13:795–825
6. Jiménez C, Antonio J (2018) Interpolación Spline y aplicación a las curvas de nivel
7. Mina-Valdés A (2011) Obtaining and projection of mortality tables using spline curves. Papeles de Población 17(69):49–72
8. Trincado G, Vidal J (1999) Aplicación de interpolación "spline" cúbica en la estimación de volumen. Bosque 20(2):3–8

9. McConnell JJ (2005) Computer graphics: theory into practice. Jones and Bartlett Publishers
10. Bercovier M, Matskewich T (2017) Smooth Bézier surfaces over unstructured quadrilateral meshes
11. Mederos Anoceto OB, González BE, Others (1997) Una variante metodológica para el estudio de los conceptos a partir de su definición. Boletín de Matemáticas 4(2):85–100
12. Gonzalez FJ, Fumeaux C, Alda J, Boreman GD (2000) Thermal impedance model of electrostatic discharge effects on microbolometers. Microw Opt Technol Lett 26(5):291–293
13. Gonzalez FJ, Gritz MA, Fumeaux C, Boreman GD (2002) Two dimensional array of antenna-coupled microbolometers. Int J Infrared Millimeter Waves 23(5):785–797
14. García-Valdivieso G, Navarro-Contreras HR, Vera-Reveles G, González FJ, Simmons TJ, Hernández MG et al (2017) High sensitivity bolometers from thymine functionalized multi-walled carbon nanotubes. Sens Actuators B Chem 238:880–887
15. Kruse PW et al (2001) Uncooled thermal imaging: arrays, systems, and applications. SPIE press Bellingham, WA
16. Zhou X-F, Zhang H, Li Y, Tang X-D, Chen Q-M, Zhang P-X (2009) Giant temperature coefficient of resistance in zno/si (111) thin films. Chin Phys Lett 27(1):18101
17. Liu B, Liu C, Xu J, Yi B (2010) Temperature coefficients of grain boundary resistance variations in a ZnO/p-Si heterojunction. J Semicond 31(12):122001
18. Wang B, Lai J, Li H, Hu H, Chen S (2013) Nanostructured vanadium oxide thin film with high TCR at room temperature for microbolometer. Infrared Phys Technol 57:8–13
19. Choi S, Kim B-J, Lee YW, Yun SJ, Kim H-T (2008) Synthesis of VO2 Nanowire and Observation of Metal-Insulator Transition. Jpn J Appl Phys 47(4S):3296
20. Zhou X (2015) Ag-doping improving the detection sensitivity of bolometer based on ZnO thin films. Vacuum 117:47–49
21. Sanchez JE, González G, Vera-Reveles G, Velazquez-Salazar JJ, Bazan-Diaz L, Gutiérrez-Hernández JM et al (2017) Silver/zinc oxide self-assembled nanostructured bolometer. Infrared Phys Technol 81:266–270
22. Sanchez JE, Mendoza-Santoyo F, Cantu-Valle J, Velazquez-Salazar J, José Yacaman M, González FJ et al (2015) Electric radiation mapping of silver/zinc oxide nanoantennas by using electron holography. J Appl Phys 117(3):34306
23. Sun Y, Gates B, Mayers B, Xia Y (2002) Crystalline silver nanowires by soft solution processing. Nano Lett 2(2):165–168
24. Sun Y, Yin Y, Mayers BT, Herricks T, Xia Y (2002) Uniform silver nanowires synthesis by reducing AgNO3 with ethylene glycol in the presence of seeds and poly (vinyl pyrrolidone). Chem Mater 14(11):4736–4745
25. Ahmed M, Chitteboyina MM, Butler DP, Celik-Butler Z (2012) Temperature sensor in a flexible substrate. IEEE Sens J 12(5):864–869
26. Atar FB, Yesilyurt A, Onbasli MC, Hanoglu O, Okyay AK (2011) Ge/SiGe Quantum well p-i-n structures for uncooled infrared bolometers. IEEE Electron Device Lett 32(11):1567–1569
27. Yan J, Kim M-H, Elle JA, Sushkov AB, Jenkins GS, Milchberg HM et al (2012) Dual-gated bilayer graphene hot-electron bolometer. Nat Nanotechnol 7(7):472–478
28. Glamazda AY, Karachevtsev VA, Euler WB, Levitsky IA (2012) Achieving high mid-IR bolometric responsivity for anisotropic composite materials from carbon nanotubes and polymers. Adv Func Mater 22(10):2177–2186
29. Santiago-Valenten E, Portilla-Flores EA, Mezura-Montes E, Vega-Alvarado E, Calva-Yanez MB, Pedroza-Villalba M (2019) A graph-theory-based method for topological and dimensional representation of planar mechanisms as a computational tool for engineering design. IEEE Access 7:587–596
30. Johansson S, Rossi M, Hall SA, Sparrenbom C, Hagerberg D, Tudisco E et al (2019) Combining spectral induced polarization with X-ray tomography to investigate the importance of DNAPL geometry in sand samples. Geophysics 84(3):E173–E188
31. Papari GP, Koral C, Andreone A (2019) Geometrical dependence on the onset of surface plasmon polaritons in THz grid metasurfaces. Sci Rep 9(1):924

32. Díaz de León-Zapata R, González G, Flores-García E, Rodríguez ÁG, González FJ (2016) Evolutionary algorithm geometry optimization of optical antennas. Int J Antennas Propag 2016:1–7
33. Diaz de Leon-Zapata R, González-Fernández JV, Flores-García E, de la Rosa-Zapata AB, Lara-Velázquez I (2018) Multi-objective evolutionary algorithm as a method to obtain optimized nanostructures. Eur Phys J Appl Phys 83(2):20502
34. Zhang JZ (2003) Self-assembled nanostructures. Kluwer Academic/Plenum Publishers
35. Sharma AK, Pandey AK, Kaur B (2018) A Review of advancements (2007–2017) in plasmonics-based optical fiber sensors. Opt Fiber Technol 43:20–34
36. Maier SA (2007) Plasmonics: fundamentals and applications. Springer
37. Kwon H, Chalabi H, Alù A (2019) Refractory Brewster metasurfaces control the frequency and angular spectrum of light absorption. Nanomater Nanotechnol 9:184798041882481
38. Whitley D (2019) Next generation genetic algorithms: a user's guide and tutorial. Springer, Cham, pp 245–274

On Complex Periodic Evolutions of a Periodically Diffused Brusselator

Siyu Guo and Albert C. J. Luo

Complex periodic evolutions in a periodically Diffused Brusselator are presented through the generalized harmonic balance method, and the corresponding stability and bifurcations of the periodic evolutions are determined through eigenvalue analysis. For a better understanding of complex periodic evolutions of chemical reactions, numerical simulations of stable periodic evolutions are completed for illustrations of complex periodic evolutions. This chapter is dedicated to Professor Valentin Afraimovich for a friendship between Valentin and Albert in the past two decades.

1 Introduction

In 1950, modern nonlinear chemical dynamics was initiated by Boris Pavlovich Belousov. Only an abstract of his work [1] was published in 1951. After 15 years, Zhabotinsky [2–4] followed the Belousov's work to systematically study chemical reaction dynamics, and the corresponding chemical reaction was called the Belousov-Zhabotinsky reaction. In 1968, Zhabotinsky presented the Belousov-Zhabotinsky reaction in a conference in Prague. To help one understand the evolution of Belousov-Zhabotinsky reaction, Field and Noyes [5] published a paper to interpret such an observation of the Belousov-Zhabotinsky reaction.

In 1968, Prigogine and Lefever [7] proposed a mathematical model for four-step chemical reaction, which was named the Brusselator in Tyson [6]. The four chemical reaction model is presented as

$$A \xrightarrow{k_1} X$$

S. Guo · A. C. J. Luo (✉)
Department of Mechanical and Mechatronics Engineering, Southern Illinois University Edwardsville, Edwardsville IL62026-1805, USA
e-mail: aluo@siue.edu

13

$$B + X \xrightarrow{k_2} Y + D$$

$$2X + Y \xrightarrow{k_3} 3X$$

$$X \xrightarrow{k_4} E \tag{1}$$

where A and B are input chemicals, X and Y are intermediate chemicals, D and E are output chemicals. $k_i (i = 1, 2, 3, 4)$ are rate constant of each sub-step respectively. For convenience, $[A]$, $[B]$, $[D]$, $[E]$, $[X]$ and $[Y]$ denote the concentrations of chemicals A, B, D, E, X and Y respectively. The capital letter "T" is time. Introduce new variables

$$x = \sqrt{\frac{k_3}{k_4}}[X], \quad y = \sqrt{\frac{k_3}{k_4}}[Y], \quad t = k_4 T$$

$$a = \frac{k_1}{k_4}\sqrt{\frac{k_3}{k_4}}[A], \quad b = \frac{k_2}{k_4}[B], \quad d = \sqrt{\frac{k_3}{k_4}}[D], \quad e = \sqrt{\frac{k_3}{k_4}}[E] \tag{2}$$

The differential equations of the Brusselator are

$$\dot{x} = a - (b + 1)x + x^2 y$$

$$\dot{y} = bx - x^2 y \tag{3}$$

where $\dot{x} = dx/dt$ and $\dot{y} = dy/dt$. In Lefever and Nicolis [8], proved such a Brusselator possesses was a unique limit cycle near an unstable equilibrium. As in Tomita et al. [9], the periodically forced Brusselator model with a harmonic diffusion is

$$\dot{x} = a - (b + 1)x + x^2 y + Q_0 \cos \Omega t$$

$$\dot{y} = bx - x^2 y \tag{4}$$

where a and b are constant because constant input concentrations are usually wanted. Q_0 and Ω are excitation amplitude and frequency, respectively. Thus, numerical and perturbation methods was used to study dynamics of the Brusselator. However, the harmonic balance method was adopted for periodic motions in mechanical and electronic problems (e.g., [10, 11]). The truncation error cannot be estimated in the traditional harmonic balance method. This is because harmonic terms in the harmonic balance method are selected randomly.

To avoid such a problem in the traditional harmonic balance method, Luo [12] proposed the generalized harmonic balance method in 2012. Such a method is a transformation based on infinite Fourier series. Such a transformation converted a dynamical system to a new dynamical system of coefficients in the Fourier series, which was like the Laplace transform of linear dynamical systems. A few classic nonlinear dynamical systems were studied through the generalized harmonic balance method (e.g., [13–20]). For the periodically forced Brusselator et al. [21] employed the generalized harmonic balance method for the analytical periodic evolutions of the periodically diffused Brusselator. Such methodology provided a good accuracy of the periodic evolutions in the periodically forced Brusselator.

In this chapter, the analytical solutions of stable and unstable periodic evolutions in the periodically diffused Brusselator will be presented through the generalized harmonic balance method. For a better understanding of chemical nonlinear oscillator, complex periodic evolutions of the periodically diffused Brusselator will be presented, and the harmonic spectrums of the periodic evolutions will be presented to show the slow and fast evolutions.

2 Analytical Solutions

Consider a nonlinear dynamical system

$$\dot{\mathbf{x}} = \mathbf{f}(\mathbf{x}, t), \tag{5}$$

where

$$\mathbf{x} = (x, y)^{\mathrm{T}}, \dot{\mathbf{x}} = d\mathbf{x}/dt, \mathbf{f} = (f_1(x, y, t), f_2(x, y, t))^{\mathrm{T}}. \tag{6}$$

For the forced Brusselator system in Eq. (4), the vector field $\mathbf{f}(\mathbf{x}, t)$ is

$$
\begin{aligned}
f_1(x, y, t) &= a - (b + 1)x + x^2 y + Q_0 \cos \Omega t, \\
f_2(x, y, t) &= bx - x^2 y.
\end{aligned}
\tag{7}
$$

As in Luo [12], a periodic solution of Eq. (4) is

$$
\begin{aligned}
x^{(m)*}(t) &\approx a_{(1)0/m} + \sum_{l=1}^{N} b_{(1)l/m} \cos(\frac{l}{m}\Omega t) + c_{(1)l/m} \sin(\frac{l}{m}\Omega t), \\
y^{(m)*}(t) &\approx a_{(2)0/m} + \sum_{l=1}^{N} b_{(2)l/m} \cos(\frac{l}{m}\Omega t) + c_{(2)l/m} \sin(\frac{l}{m}\Omega t).
\end{aligned}
\tag{8}
$$

Taking derivative of Eq. (8) with respect to time t gives

$$
\begin{aligned}
\dot{x}^{(m)*}(t) &\approx \dot{a}_{(1)0/m} + \sum_{l=1}^{N} (\dot{b}_{(1)l/m} + \frac{l}{m}\Omega c_{(1)l/m}) \cos(\frac{l}{m}\Omega t) \\
&\quad + \sum_{l=1}^{N} (\dot{c}_{(1)l/m} - \frac{l}{m}\Omega b_{(1)l/m}) \sin(\frac{l}{m}\Omega t),
\end{aligned}
$$

$$\dot{y}^{(m)*}(t) \approx \dot{a}_{(2)0/m} + \sum_{l=1}^{N} \left(\dot{b}_{(2)l/m} + \frac{l}{m}\Omega c_{(2)l/m} \right) \cos\left(\frac{l}{m}\Omega t\right)$$

$$+ \sum_{l=1}^{N} \left(\dot{c}_{(2)l/m} - \frac{l}{m}\Omega b_{(2)l/m} \right) \sin\left(\frac{l}{m}\Omega t\right). \tag{9}$$

Substitution of Eqs. (8) and (9) into Eq. (5) with averaging produces

$$\dot{a}_{(1)0/m} = F_{(1)0/m}(\mathbf{z}^{(m)}),$$

$$\dot{b}_{(1)l/m} = -\frac{l}{m}\Omega c_{(1)l/m} + F_{(1)l/m}^{(c)}(\mathbf{z}^{(m)}) \quad (l = 1, 2, \ldots, N),$$

$$\dot{c}_{(1)l/m} = \frac{l}{m}\Omega b_{(1)l/m} + F_{(1)l/m}^{(s)}(\mathbf{z}^{(m)}) \quad (l = 1, 2, \ldots, N);$$

$$\dot{a}_{(2)0/m} = F_{(2)0/m}(\mathbf{z}^{(m)}),$$

$$\dot{b}_{(2)l/m} = -\frac{l}{m}\Omega c_{(2)l/m} + F_{(2)l/m}^{(c)}(\mathbf{z}^{(m)}) \quad (l = 1, 2, \ldots, N),$$

$$\dot{c}_{(2)l/m} = \frac{l}{m}\Omega b_{(2)l/m} + F_{(2)l/m}^{(s)}(\mathbf{z}^{(m)}) \quad (l = 1, 2, \ldots, N), \tag{10}$$

where

$$F_{(1)0/m}(\mathbf{z}^{(m)}) = \frac{1}{mT} \int_0^{mT} f_1(x^{(m)*}, y^{(m)*}, t)dt,$$

$$F_{(1)l/m}^{(c)}(\mathbf{z}^{(m)}) = \frac{2}{mT} \int_0^{mT} f_1(x^{(m)*}, y^{(m)*}, t) \cos\left(\frac{l}{m}\Omega t\right)dt,$$

$$F_{(1)l/m}^{(s)}(\mathbf{z}^{(m)}) = \frac{2}{mT} \int_0^{mT} f_1(x^{(m)*}, y^{(m)*}, t) \sin\left(\frac{l}{m}\Omega t\right)dt;$$

$$F_{(2)0/m}(\mathbf{z}^{(m)}) = \frac{1}{mT} \int_0^{mT} f_2(x^{(m)*}, y^{(m)*}, t)dt,$$

$$F_{(2)l/m}^{(c)}(\mathbf{z}^{(m)}) = \frac{2}{mT} \int_0^{mT} f_2(x^{(m)*}, y^{(m)*}, t) \cos\left(\frac{l}{m}\Omega t\right)dt,$$

$$F_{(2)l/m}^{(s)}(\mathbf{z}^{(m)}) = \frac{2}{mT} \int_0^{mT} f_2(x^{(m)*}, y^{(m)*}, t) \sin\left(\frac{l}{m}\Omega t\right)dt. \tag{11}$$

The constant term for concentration x is

$$F_{(1)0/m} = a - (b+1)a_{(1)0/m} + f_{0/m}^{(1)} + \frac{1}{2}\sum_{k=1}^{N} f_{0/m}^{(2)} + \frac{1}{4}\sum_{i=1}^{N}\sum_{j=1}^{N}\sum_{k=1}^{N} f_{0/m}^{(3)}. \quad (12)$$

The cosine term for concentration x is

$$F_{(1)l/m}^{(c)} = -(b+1)b_{(1)l/m} + Q_0\delta_m^l + f_{l/m}^{(c,1)} + \frac{1}{2}\sum_{i=1}^{N}\sum_{j=1}^{N} f_{l/m}^{(c,2)}$$

$$+ \frac{1}{4}\sum_{i=1}^{N}\sum_{j=1}^{N}\sum_{k=1}^{N} f_{l/m}^{(c,3)}. \quad (13)$$

The sine term for the concentration x is

$$F_{(1)l/m}^{(s)} = -(b+1)c_{(1)l/m} + f_{l/m}^{(s,1)} + \frac{1}{2}\sum_{i=1}^{N}\sum_{j=1}^{N} f_{l/m}^{(s,2)} + \frac{1}{4}\sum_{i=1}^{N}\sum_{j=1}^{N}\sum_{k=1}^{N} f_{l/m}^{(s,3)}. \quad (14)$$

The constant term for the concentration y is

$$F_{(2)0/m} = ba_{(1)0/m} - f_{0/m}^{(1)} - \frac{1}{2}\sum_{k=1}^{N} f_{0/m}^{(2)} - \frac{1}{4}\sum_{i=1}^{N}\sum_{j=1}^{N}\sum_{k=1}^{N} f_{0/m}^{(3)}. \quad (15)$$

The cosine term for the concentration y is

$$F_{(2)l/m}^{(c)} = bb_{(1)l/m} - f_{l/m1}^{(c,1)} - \frac{1}{2}\sum_{i=1}^{N}\sum_{j=1}^{N} f_{l/m}^{(c,2)} - \frac{1}{4}\sum_{i=1}^{N}\sum_{j=1}^{N}\sum_{k=1}^{N} f_{l/m}^{(c,3)}. \quad (16)$$

The sine term for the concentration y is

$$F_{(2)l/m}^{(s)} = bc_{(1)l/m} - f_{l/m}^{(s,1)} - \frac{1}{2}\sum_{i=1}^{N}\sum_{j=1}^{N} f_{l/m}^{(s,2)} - \frac{1}{4}\sum_{i=1}^{N}\sum_{j=1}^{N}\sum_{k=1}^{N} f_{l/m}^{(s,3)}. \quad (17)$$

The corresponding functions are

$$f_{0/m}^{(1)} = (a_{(1)0/m})^2 a_{(2)0/m},$$
$$f_{0/m}^{(2)} = a_{(2)0/m}(b_{(1)k/m}b_{(1)k/m} + c_{(1)k/m}c_{(1)k/m})$$
$$+ 2a_{(1)0/m}(b_{(1)k/m}b_{(2)k/m} + c_{(1)k/m}c_{(2)k/m}),$$
$$f_{0/m}^{(3)} = b_{(1)i/m}b_{(1)j/m}b_{(2)k/m}\Delta_1 + (2b_{(1)i/m}c_{(1)j/m}c_{(1)k/m} + b_{(2)i/m}c_{(1)j/m}c_{(1)k/m})\Delta_2,$$
$$f_{l/m}^{(c,1)} = (a_{(1)0/m})^2 b_{(2)l/m},$$

$$f_{l/m}^{(c,2)} = (a_{(2)0/m}b_{(1)i/m}b_{(1)j/m} + 2a_{(1)0/m}b_{(1)i/m}b_{(2)j/m})\Delta_{11}$$
$$+ (a_{(2)0/m}c_{(1)i/m}c_{(1)j/m} + 2a_{(1)0/m}c_{(1)i/m}c_{(2)j/m})\Delta_{12},$$
$$f_{l/m}^{(c,3)} = b_{(1)i/m}b_{(1)j/m}b_{(2)k/m}\Delta_{13} + (2b_{(1)i/m}c_{(1)j/m}c_{(2)k/m} + b_{(2)i/m}c_{(1)j/m}c_{(1)k/m})\Delta_{14},$$
$$f_{l/m}^{(s,1)} = 2a_{(1)0/m}a_{(2)0/m}c_{(1)l/m} + (a_{(1)0/m})^2 c_{(2)l/m},$$
$$f_{l/m}^{(s,2)} = (a_{(2)0/m}b_{(1)i/m}c_{(1)j/m} + a_{(1)0/m}b_{(2)i/m}c_{(1)j/m} + a_{(1)0/m}b_{(1)i/m}c_{(2)j/m})\Delta_{21},$$
$$f_{l/m}^{(s,3)} = (2b_{(1)i/m}b_{(2)j/m}c_{(1)k/m} + b_{(1)i/m}b_{(1)j/m}c_{(2)k/m})\Delta_{22} + c_{(1)i/m}c_{(1)j/m}c_{(2)k/m}\Delta_{23},$$

$$(18)$$

where

$$\Delta_1 = \delta_{i+j-k}^0 + \delta_{i-j+k}^0 + \delta_{i-j-k}^0,$$
$$\Delta_2 = \delta_{i+j-k}^0 + \delta_{i-j+k}^0 - \delta_{i-j-k}^0,$$
$$\Delta_{11} = \delta_{i+j}^l + \delta_{|i-j|}^l, \quad \Delta_{12} = -\delta_{i+j}^l + \delta_{|i-j|}^l,$$
$$\Delta_{13} = \delta_{i+j+k}^l + \delta_{|i+j-k|}^l + \delta_{|i-j+k|}^l + \delta_{|i-j-k|}^l,$$
$$\Delta_{14} = \delta_{|i+j-k|}^l - \delta_{i+j+k}^l + \delta_{|i-j+k|}^l - \delta_{|i-j-k|}^l,$$
$$\Delta_{21} = \delta_{i+j}^l - \mathrm{sgn}(i-j)\delta_{|i-j|}^l,$$
$$\Delta_{22} = \delta_{i+j+k}^l - \mathrm{sgn}(i+j-k)\delta_{|i+j-k|}^l + \mathrm{sgn}(i-j+k)\delta_{|i-j+k|}^l$$
$$- \mathrm{sgn}(i-j-k)\delta_{|i-j-k|}^l,$$
$$\Delta_{23} = -\delta_{i+j+k}^l + \mathrm{sgn}(i+j-k)\delta_{|i+j-k|}^l + \mathrm{sgn}(i-j+k)\delta_{|i-j+k|}^l$$
$$- \mathrm{sgn}(i-j-k)\delta_{|i-j-k|}^l.$$

$$(19)$$

In Eqs. (10) and (11), the vector $\mathbf{z}^{(m)}$ is constructed by concatenating constants, sine term coefficients and cosine term coefficients in Eq. (8), i.e.,

$$\mathbf{z}^{(m)} \equiv (a_{(1)0/m}, \mathbf{b}_{(1)m}^T, \mathbf{c}_{(1)m}^T, a_{(2)0/m}, \mathbf{b}_{(2)m}^T, \mathbf{c}_{(2)m}^T)^T$$
$$= (z_1, z_2, \ldots, z_{2N+1}, z_{2N+2}, \ldots, z_{4N+2})^T \qquad (20)$$

where

$$\mathbf{b}_{(1)m} \equiv (b_{(1)1/m}, \ldots, b_{(1)N/m})^T,$$
$$\mathbf{c}_{(1)m} \equiv (c_{(1)1/m}, \ldots, c_{(1)N/m})^T,$$
$$\mathbf{b}_{(2)m} \equiv (b_{(2)1/m}, \ldots, b_{(2)N/m})^T,$$
$$\mathbf{c}_{(2)m} \equiv (c_{(2)1/m}, \ldots, c_{(2)N/m})^T. \qquad (21)$$

Equation (10) is rewritten in a standard form as

$$\dot{\mathbf{z}}^{(m)} = \mathbf{g}^{(m)}(\mathbf{z}^{(m)}), \tag{22}$$

where

$$\mathbf{g}^{(m)}(\mathbf{z}^{(m)}) = \begin{Bmatrix} F_{(1)0/m} \\ -\dfrac{\Omega}{m}\mathbf{k}\mathbf{c}_{(1)m} + \mathbf{F}_{(1)}^{(c)} \\ \dfrac{\Omega}{m}\mathbf{k}\mathbf{b}_{(1)m} + \mathbf{F}_{(1)}^{(s)} \\ F_{(2)0/m} \\ -\dfrac{\Omega}{m}\mathbf{k}\mathbf{c}_{(2)m} + \mathbf{F}_{(2)}^{(c)} \\ \dfrac{\Omega}{m}\mathbf{k}\mathbf{b}_{(2)m} + \mathbf{F}_{(2)}^{(s)} \end{Bmatrix}, \tag{22}$$

and

$$\mathbf{k} = \mathrm{diag}(1, 2, \ldots, N),$$
$$\mathbf{F}_{(1)}^{(c)} = (F_{(1)1/m}^{(c)}, F_{(1)2/m}^{(c)}, \ldots, F_{(1)N/m}^{(c)})^{\mathrm{T}},$$
$$\mathbf{F}_{(1)}^{(s)} = (F_{(1)1/m}^{(s)}, F_{(1)2/m}^{(s)}, \ldots, F_{(1)N/m}^{(s)})^{\mathrm{T}},$$
$$\mathbf{F}_{(2)}^{(c)} = (F_{(2)1/m}^{(c)}, F_{(2)2/m}^{(c)}, \ldots, F_{(2)N/m}^{(c)})^{\mathrm{T}},$$
$$\mathbf{F}_{(2)}^{(s)} = (F_{(2)1/m}^{(s)}, F_{(2)2/m}^{(s)}, \ldots, F_{(2)N/m}^{(s)})^{\mathrm{T}}. \tag{24}$$

If the periodic solution of system in Eq. (5) exists, then constants and harmonic coefficients in Eq. (8) are constant, which means $\dot{\mathbf{z}}^{(m)} = \mathbf{0}$ in Eq. (22), i.e.,

$$F_{(1)0/m} = 0,$$
$$-\dfrac{\Omega}{m}\mathbf{k}\mathbf{c}_{(1)m} + \mathbf{F}_{(1)}^{(c)} = \mathbf{0},$$
$$\dfrac{\Omega}{m}\mathbf{k}\mathbf{b}_{(1)m} + \mathbf{F}_{(1)}^{(s)} = \mathbf{0};$$
$$F_{(2)0/m} = 0,$$
$$-\dfrac{\Omega}{m}\mathbf{k}\mathbf{c}_{(2)m} + \mathbf{F}_{(2)}^{(c)} = \mathbf{0},$$
$$\dfrac{\Omega}{m}\mathbf{k}\mathbf{b}_{(2)m} + \mathbf{F}_{(2)}^{(s)} = \mathbf{0}. \tag{25}$$

The $2(2N + 1)$-algebraic equations in Eq. (25) can be solved by the Newton–Raphson algorithm, and the analytical solution of Eq. (25) is for the truncated Fourier series coefficients in Eq. (8).

The stability of such a periodic solution is determined through the equilibrium of the transformed system. The linearization of the transformed system at the equilibrium $\mathbf{z}^{*(m)}$ is

$$\Delta\dot{\mathbf{z}}^{(m)} = D\mathbf{g}^{(m)}(\mathbf{z}^{*(m)})\Delta\mathbf{z}^{(m)} = \left.\frac{\partial\mathbf{g}^{(m)}(\mathbf{z}^{(m)})}{\partial\mathbf{z}^{(m)}}\right|_{\mathbf{z}^{(m)}=\mathbf{z}^{*(m)}} \cdot \Delta\mathbf{z}^{(m)}, \qquad (26)$$

where

$$D\mathbf{g}^{(m)}(\mathbf{z}^{*(m)}) = \frac{\partial\mathbf{g}^{(m)}}{\partial\mathbf{z}^{(m)}} = \begin{bmatrix} \frac{\partial g_1}{\partial z_1} & \frac{\partial g_1}{\partial z_2} & \cdots & \frac{\partial g_1}{\partial z_{4N+2}} \\ \frac{\partial g_2}{\partial z_1} & \frac{\partial g_2}{\partial z_2} & \cdots & \frac{\partial g_2}{\partial z_{4N+2}} \\ \vdots & \vdots & & \vdots \\ \frac{\partial g_{4N+2}}{\partial z_1} & \frac{\partial g_{4N+2}}{\partial z_2} & \cdots & \frac{\partial g_{4N+2}}{\partial z_{4N+2}} \end{bmatrix}_{2(2N+1)\times 2(2N+1)} \qquad (27)$$

and

$$\frac{\partial g_1}{\partial z_r} = \frac{\partial F_{(1)0}(\mathbf{z}^{(m)})}{\partial z_r}, \quad \frac{\partial g_{l+1}}{\partial z_r} = \frac{\partial F_{(1)l/m}^{(c)}(\mathbf{z}^{(m)})}{\partial z_r},$$

$$\frac{\partial g_{N+1+l}}{\partial z_r} = \frac{\partial F_{(1)l/m}^{(s)}(\mathbf{z}^{(m)})}{\partial z_r}, \quad \frac{\partial g_{2N+2}}{\partial z_r} = \frac{\partial F_{(2)0}(\mathbf{z}^{(m)})}{\partial z_r},$$

$$\frac{\partial g_{2N+2+l}}{\partial z_r} = \frac{\partial F_{(2)l/m}^{(c)}(\mathbf{z}^{(m)})}{\partial z_r}, \quad \frac{\partial g_{3N+2+l}}{\partial z_r} = \frac{\partial F_{(2)l/m}^{(s)}(\mathbf{z}^{(m)})}{\partial z_r}, \qquad (28)$$

and

$$\frac{\partial g_1}{\partial z_r} = (b+1)\delta_0^{r-1} + g_{(0)r}^{(1)} + \frac{1}{2}\sum_{k=1}^{N} g_{(0)r}^{(2)} + \frac{1}{4}\sum_{i=1}^{N}\sum_{j=1}^{N}\sum_{k=1}^{N} g_{(0)r}^{(3)},$$

$$\frac{\partial g_{l+1}}{\partial z_r} = -\frac{l\Omega}{m}\delta_{l+N}^{r-1} - (b+1)\delta_l^{r-1} + g_{l/r}^{(c,1)}$$

$$+ \frac{1}{2}\sum_{i=1}^{N}\sum_{j=1}^{N} g_{l/r}^{(c,2)} + \frac{1}{4}\sum_{i=1}^{N}\sum_{j=1}^{N}\sum_{k=1}^{N} g_{l/r}^{(c,3)},$$

$$\frac{\partial g_{N+1+l}}{\partial z_r} = \frac{l\Omega}{m}\delta_l^{r-1} - (b+1)\delta_{l+3N+1}^{r-1} + g_{l/r}^{(s,1)} + \frac{1}{2}\sum_{i=1}^{N}\sum_{j=1}^{N} g_{l/r}^{(s,2)}$$

$$+ \frac{1}{4}\sum_{i=1}^{N}\sum_{j=1}^{N}\sum_{k=1}^{N} g_{l/r}^{(s,3)};$$

$$\frac{\partial g_{2N+2}}{\partial z_r} = b\delta_0^{r-1} - g_{(0)r}^{(1)} - \frac{1}{2}\sum_{k=1}^{N} g_{(0)r}^{(2)} - \frac{1}{4}\sum_{i=1}^{N}\sum_{j=1}^{N}\sum_{k=1}^{N} g_{(0)r}^{(3)},$$

$$\frac{\partial g_{2N+2+l}}{\partial z_r} = -\frac{l\Omega}{m}\delta_{l+3N+1}^{r-1} + b\delta_l^{r-1} - g_{l/r}^{(c,1)} - \frac{1}{2}\sum_{i=1}^{N}\sum_{j=1}^{N} g_{l/r}^{(c,2)}$$

$$- \frac{1}{4}\sum_{i=1}^{N}\sum_{j=1}^{N}\sum_{k=1}^{N} g_{l/r}^{(c,3)},$$

$$\frac{\partial g_{3N+2+l}}{\partial z_r} = \frac{l\Omega}{m}\delta_{l+2N+1}^{r-1} + b\delta_{l+N}^{r} - g_{l/r}^{(s,1)} - \frac{1}{2}\sum_{i=1}^{N}\sum_{j=1}^{N} g_{l/r}^{(s,2)}$$

$$- \frac{1}{4}\sum_{i=1}^{N}\sum_{j=1}^{N}\sum_{k=1}^{N} g_{l/r}^{(s,3)}, \tag{29}$$

where

$$g_{(0)r}^{(1)} = 2\delta_0^{r-1} a_{(1)0/m} a_{(2)0/m} + \delta_{2N+1}^{r-1}(a_{(1)0/m})^2 a_{(2)0/m},$$

$$g_{(0)r}^{(2)} = \delta_{2N+1}^{r-1}(b_{(1)k/m}b_{(1)k/m} + c_{(1)k/m}c_{(1)k/m})$$
$$+ 2a_{(2)0/m}(\delta_k^{r-1}b_{(1)k/m} + \delta_{k+N}^{r-1}c_{(1)k/m})$$
$$+ 2\delta_0^{r-1}(b_{(1)k/m}b_{(2)k/m} + c_{(1)k/m}c_{(2)k/m})$$
$$+ 2a_{(1)0}^{(m)}[\delta_k^{r-1}b_{(2)k/m} + \delta_{k+2N+1}^{r-1}b_{(1)k/m}$$
$$+ \delta_{k+N}^{r-1}c_{(2)k/m} + \delta_{k+3N+1}^{r-1}c_{(1)k/m}],$$

$$g_{(0)r}^{(3)} = [\delta_i^{r-1}b_{(1)j/m}b_{(2)k/m} + \delta_j^{r-1}b_{(1)i/m}b_{(2)k/m}$$
$$+ \delta_{k+2N+1}^{r-1}b_{(1)i/m}b_{(1)j/m}]\Delta_1 + [2(\delta_i^{r-1}c_{(1)j/m}c_{(1)k/m}$$
$$+ \delta_{j+N}^{r-1}b_{(1)i/m}c_{(1)k/m} + \delta_{k+N}^{r-1}b_{(1)i/m}c_{(1)j/m})$$
$$+ \delta_{i+2N+1}^{r-1}c_{(1)j/m}c_{(1)k/m} + \delta_{j+N}^{r-1}b_{(2)i/m}c_{(1)k/m}$$
$$+ \delta_{k+N}^{r-1}b_{(2)i/m}c_{(1)j/m}]\Delta_2; \tag{30}$$

$$g_{l/r}^{(c,1)} = 2\delta_0^{r-1}a_{(1)0/m}b_{(2)l/m} + \delta_{l+2N+1}^{r-1}(a_{(1)0/m})^2,$$

$$g_{l/r}^{(c,2)} = [(\delta_{2N+1}^{r-1}b_{(1)i/m}b_{(1)j/m} + \delta_i^{r-1}a_{(1)0/m}b_{(1)j/m} + \delta_j^{r-1}a_{(2)0/m}b_{(1)i/m})$$
$$+ 2(\delta_0^{r-1}b_{(1)i/m}b_{(2)j/m} + \delta_i^{r-1}a_{(1)0/m}b_{(2)j/m} + \delta_{j+2N+1}^{r-1}a_{(1)0/m}b_{(1)i/m})]\Delta_{11}$$
$$+ [(\delta_{2N+1}^{r-1}c_{(1)i/m}c_{(1)j/m} + \delta_{i+N}^{r-1}a_{(2)0/m}c_{(1)j/m} + \delta_{j+N}^{r-1}a_{(2)0/m}c_{(1)j/m})$$
$$+ 2(\delta_0^{r-1}c_{(1)i/m}c_{(2)j/m} + \delta_{i+N}^{r-1}a_{(1)0/m}c_{(2)j/m} + \delta_{j+3N+1}^{r-1}a_{(1)0/m}c_{(1)i/m})]\Delta_{12},$$

$$g_{l/r}^{(c,3)} = (\delta_i^{r-1}b_{(1)j/m}b_{(2)k/m} + \delta_j^{r-1}b_{(1)i/m}b_{(2)k/m} + \delta_{k+2N+1}^{r-1}b_{(1)i/m}b_{(1)j/m})\Delta_{13}$$
$$+ [2(\delta_i^{r-1}c_{(1)j/m}c_{(2)k/m} + \delta_{j+N}^{r-1}b_{(1)i/m}c_{(2)k/m} + \delta_{k+3N+1}^{r-1}b_{(1)i/m}c_{(1)j/m})$$

$$+ \delta_{i+2N+1}^{r-1} c_{(1)j/m} c_{(1)k/m} + \delta_{j+N}^{r-1} b_{(2)i/m} c_{(1)k/m} + \delta_{k+N}^{r-1} b_{(2)i/m} c_{(1)j/m}]\Delta_{14},$$

$$g_{l/r}^{(s,1)} = 2(\delta_0^{r-1} a_{(2)0/m} c_{(1)l/m} + \delta_{2N+1}^{r-1} a_{(1)0/m} c_{(1)l/m} + \delta_{l+N}^{r-1} a_{(1)0/m} a_{(2)0/m})$$
$$+ 2\delta_0^{r-1} a_{(1)0/m} c_{(2)l/m} + \delta_{l+3N+1}^{r-1} (a_{(1)0/m})^2,$$

$$g_{l/r}^{(s,2)} = [(\delta_{2N+1}^{r-1} b_{(1)i/m} c_{(1)j/m} + \delta_i^{r-1} a_{(2)0/m} c_{(1)j/m} + \delta_{j+N}^{r-1} a_{(2)0/m} b_{(1)i/m})$$
$$+ (\delta_0^{r-1} b_{(2)i/m} c_{(1)j/m} + \delta_{i+2N+1}^{r-1} a_{(1)0/m} c_{(1)j/m} + \delta_{j+N}^{r-1} a_{(1)0}^{(m)} b_{(2)i/m})$$
$$+ (\delta_0^{r-1} b_{(1)i/m} c_{(2)j/m} + \delta_i^{r-1} a_{(1)0/m} c_{(2)j/m} + \delta_{j+3N+1}^{r-1} a_{(1)0/m} b_{(1)i/m})]\Delta_{21},$$

$$g_{l/r}^{(s,3)} = [2(\delta_i^{r-1} b_{(2)j/m} c_{(1)k/m} + \delta_{j+2N+1}^{r-1} b_{(1)i/m} c_{(1)k/m} + \delta_{k+N}^{r-1} b_{(1)i/m} b_{(2)j/m})$$
$$+ (\delta_i^{r-1} b_{(1)j/m} c_{(2)k/m} + \delta_j^{r-1} b_{(1)i/m} c_{(2)k/m} + \delta_{k+3N+1}^{r-1} b_{(1)i/m} b_{(1)j/m})]\Delta_{22}$$
$$+ [\delta_{i+N}^{r-1} c_{(1)j/m} c_{(2)k/m} + \delta_{j+N}^{r-1} c_{(1)i/m} c_{(2)k/m} + \delta_{k+3N+1}^{r-1} c_{(1)i/m} c_{(1)j/m}]\Delta_{23}. \quad (31)$$

Once the transformed system is linearized at equilibriums, the stability of periodic solutions of the periodically forced Brusselator is determined through the eigenvalue analysis of the linearized system, i.e.,

$$\left| D\mathbf{g}^{(m)}(\mathbf{z}^{*(m)}) - \lambda \mathbf{I} \right| = 0 \quad (32)$$

(i)   If $\text{Re}\lambda_i < 0$ for $i = 1, 2, \ldots, 4N + 2$, the period-m solution is stable.
(ii)   If $\text{Re}\lambda_i > 0$ for $i \in \{1, 2, \ldots, 4N + 2\}$, the period-$m$ solution is unstable.
(iii) The boundary between stable and unstable evolutions with higher order singularity gives the bifurcation conditions and stability with higher order singularity.

Once the number of harmonic terms is determined, harmonic amplitudes varying with the diffusion frequency presented in Eq. (8) are

$$x^{(m)*}(t) \approx a_{(1)0/m} + \sum_{l=1}^{N} A_{(1)l/m} \sin(\frac{l}{m}\Omega t + \varphi_{(1)l/m}),$$
$$y^{(m)*}(t) \approx a_{(2)0/m} + \sum_{l=1}^{N} A_{(2)l/m} \sin(\frac{l}{m}\Omega t + \varphi_{(2)l/m}), \quad (33)$$

where

$$A_{(1)l/m} = \sqrt{b_{(1)l/m}^2 + c_{(1)l/m}^2}, \varphi_{(1)l/m} = \arctan \frac{b_{(1)l/m}}{c_{(1)l/m}};$$
$$A_{(2)l/m} = \sqrt{b_{(2)l/m}^2 + c_{(2)l/m}^2}, \varphi_{(2)l/m} = \arctan \frac{b_{(2)l/m}}{c_{(2)l/m}}. \quad (34)$$

3 Periodic Evolutions on the Bifurcation Tree

To better understand the periodic evolutions in the periodically forced Brusselator, numerical simulation is performed by the mid-point scheme. Initial conditions in the numerical simulations are obtained from analytical solution by setting $t = 0$ in the truncated Fourier expansion in Eq. (8). Consider system parameters in Eq. (4) as

$$a = 0.4, \ b = 1.2, \ Q_0 = 0.08. \tag{35}$$

The numerical illustrations will be completed.

In the following discussion, circular symbols give the analytical solution, and solid curves represent numerical simulation results. The acronym "I.C." with a large circle is used for initial condition. The initial conditions for stable periodic evolutions are given in Table 1.

In Fig. 1, the orbits of both concentrations x and y of the different period-m evolutions ($m = 1, 2, 4, 8$) are presented. For comparison, all the concentration orbits are in the same scale. In Fig. 1i, the concentration orbit of the period-1 evolution is a smooth cycle like a harmonic oscillation. The numerical and analytical results match very well. For the period-1 evolution at $\Omega = 1.05$, 12 harmonic terms for both concentration of x and y are presented in Fig. 1ii, iii. For the concentration x, the constant term is $a_{(1)0} = 0.4$ and the main harmonic amplitude is $A_{(1)1} \approx 0.0890$. The other amplitudes are $A_{(1)l} \in (10^{-15}, 10^{-2})$ ($l = 2, 3, \ldots, 12$) with $A_{(1)12} \approx 8.6 \times 10^{-15}$. For the concentration y, the constant term is $a_{(2)0} \approx 2.9304$ and the main harmonic amplitude is $A_{(2)1} \approx 0.0960$. The other amplitudes are $A_{(2)l} \in (10^{-15}, 10^{-2})$ ($l = 2, 3, \ldots, 12$) with $A_{(2)12} \approx 8.6 \times 10^{-15}$. For such a period-1 evolution, one harmonic term with $A_{(1)1} \approx 0.0890$ and $A_{(2)1} \approx 0.0960$ can give an approximate solution. The analytical solution with 12 harmonic terms gives the period-1 evolutions with accuracy of $\varepsilon = 10^{-15}$.

The concentration orbit of the period-2 evolution with diffusion frequency $\Omega = 0.95$ is presented in Fig. 2i. A slow varying segment can be observed near the initial condition represented by circular symbols. Such an orbit of the period-2 evolution is a fast-slow varying evolution, and the analytical and numerical results also match very well. In Fig. 2ii, iii, spectrums of concentrations x and y are presented. For the concentration x, the constant term is $a_{(1)0} = 0.4$, and the main harmonic amplitudes are $A_{(1)1/2} \approx 0.1456$, $A_{(1)1} \approx 0.0816$, $A_{(1)3/2} \approx 0.0258$. The other amplitudes are $A_{(1)l/2} \in (10^{-15}, 10^{-2})$ ($l = 4, 5, \ldots, 30$) with $A_{(1)15} \approx 1.5 \times 10^{-15}$. For

Table 1 Input data for numerical simulations ($a = 0.4, b = 1.2, Q_0 = 0.08$)	Ω	Initial conditions	P-m evolution
	1.05	$(x_0, y_0) \approx (0.387807, 3.023962)$	P-1 (stable)
	0.95	$(x_0, y_0) \approx (0.368464, 2.625488)$	P-2 (stable)
	0.93	$(x_0, y_0) \approx (0.341157, 2.472344)$	P-4 (stable)
	0.92	$(x_0, y_0) \approx (0.334620, 2.462824)$	P-8 (stable)

Fig. 1 Stable period-1 evolutions ($\Omega = 1.05$): (i) concentration orbit (x, y), (ii) harmonic spectrum for concentration x, (iii) harmonic spectrum for concentration y. ($a = 0.4$, $b = 1.2$, $Q_0 = 0.08$), IC: $(x_0, y_0) \approx$ (0.387807, 3.023962). The solid curves and circular symbols are for numerical and analytical results

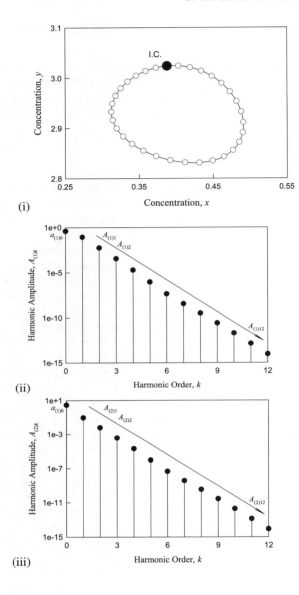

(i)

(ii)

(iii)

the concentration y, the constant term is $a_{(2)0} \approx 2.8054$, and the main harmonic amplitudes are $A_{(2)1/2} \approx 0.3393$, $A_{(2)1} \approx 0.1175$, $A_{(2)3/2} \approx 0.0315$. The other amplitudes are $A_{(2)l/2} \in (10^{-15}, 10^{-2})$ $(l = 4, 5, \cdots, 15)$ with $A_{(2)15} \approx 1.5 \times 10^{-15}$. For such a period-2 evolution, three harmonic terms can give an approximate solution. The analytical solution with 30 harmonic terms gives the period-2 evolutions with accuracy of $\varepsilon = 10^{-15}$. The main harmonic effects on the period-2 evolution are from $A_{(1)1/2} \approx 0.1456$, and $A_{(2)1/2} \approx 0.3393$.

Fig. 2 Stable period-2 evolutions ($\Omega = 0.95$): (i) concentration orbit (x, y), (ii) harmonic spectrum for concentration x, (iii) harmonic spectrum for concentration y.($a = 0.4$, $b = 1.2$, $Q_0 = 0.08$), IC: $(x_0, y_0) \approx (0.368464, 2.625488)$. The solid curves and circular symbols are for numerical and analytical results

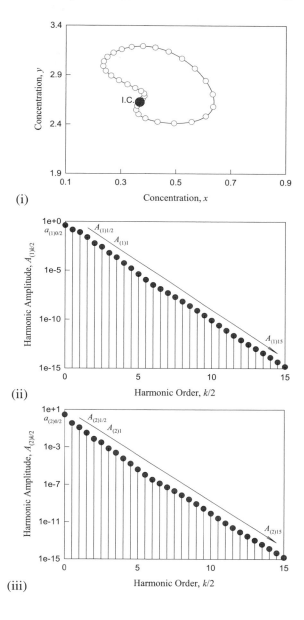

(i)

(ii)

(iii)

In Fig. 3i, the concentration orbit for the period-4 evolution is presented with $\Omega = 0.93$. The concentration orbit has two cycles which are the doubled period-2 evolution. In Fig. 3ii, iii, spectrums of concentrations x and y are presented. For the concentration x, the constant term is $a_{(1)0} = 0.4$, and the main harmonic amplitudes are $A_{(1)1/4} \approx 0.0293, A_{(1)1/2} \approx 0.1521, A_{(1)3/4} \approx 0.0265,\ A_{(1)1} \approx 0.0886,\ A_{(1)5/4} \approx$

Fig. 3 Stable period-4
evolutions ($\Omega = 0.93$): (i)
concentration orbit (x, y), (ii)
harmonic spectrum for
concentration x, (iii)
harmonic spectrum for
concentration y.($a =$
0.4, $b = 1.2$, $Q_0 = 0.08$),
IC: $(x_0, y_0) \approx$
(0.341157, 2.472344). The
solid curves and circular
symbols are for numerical
and analytical results

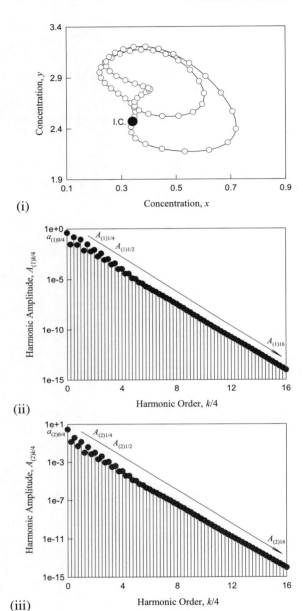

0.0293, $A_{(1)3/2} \approx 0.0816$. The other amplitudes are $A_{(1)l/4} \in (10^{-15}, 10^{-2})$ ($l = 7, 8, \ldots, 64$) with $A_{(1)16} \approx 9.5 \times 10^{-15}$. For the concentration y, the constant term is $a_{(2)0} \approx 2.7818$, and the main harmonic amplitudes are $A_{(2)1/4} \approx 0.1293$, $A_{(2)1/2} \approx 0.3608$, $A_{(2)3/4} \approx 0.0464$, $A_{(2)1} \approx 0.1309$, $A_{(2)5/4} \approx 9.5880 \times 10^{-3}$, $A_{(2)3/2} \approx 0.0364$. The other amplitudes are $A_{(2)l/2} \in (10^{-15}, 10^{-2})$ ($l = 7, 8, \ldots, 64$) with $A_{(2)16} \approx 9.5 \times 10^{-15}$. For such a period-4 evolution, one harmonic term can give an approximate solution. The analytical solution with 48 harmonic terms gives the period-4 evolutions with accuracy of $\varepsilon = 10^{-15}$. The main harmonic effects on the period-4 evolution are from $A_{(1)1/2} \approx 0.1521$ and $A_{(2)1/2} \approx 0.3608$.

In Fig. 4i, the concentration orbit for the period-8 evolution has four cycles, doubling cycles of the period-4 evolution. The period-8 evolution becomes more complicated, compared to the period-4 evolutions. In Fig. 4ii, iii, spectrums of concentrations x and y are presented. For the concentration x, the constant term is $a_{(1)0} = 0.4$, and the main harmonic amplitudes are $A_{(1)1/8} \approx 2.4187 \times 10^{-3}$, $A_{(1)1/4} \approx 0.0346$, $A_{(1)3/8} \approx 8.8178 \times 10^{-3}$, $A_{(1)1/2} \approx 0.1551, A_{(1)5/8} \approx 7.0219 \times 10^{-3}$, $A_{(1)3/4} \approx 0.0313$, $A_{(1)7/8} \approx 3.8717 \times 10^{-3}$, $A_{(1)1} \approx 0.0925$, $A_{(1)9/8} \approx 1.8867 \times 10^{-3}$, $A_{(1)5/4} \approx 9.2403 \times 10^{-3}$, $A_{(1)11/8} \approx 1.8809 \times 10^{-3}$, $A_{(1)3/2} \approx 0.0317$. The other amplitudes are $A_{(1)l/8} \in (10^{-15}, 10^{-2})$ ($l = 13, 14, \ldots, 128$) with $A_{(1)16} \approx 4.38 \times 10^{-14}$. For the concentration y, the constant term is $a_{(2)0} \approx 2.7701$, and the main harmonic amplitudes are $A_{(2)1/8} \approx 0.0212$, $A_{(2)1/4} \approx 0.1542$, $A_{(2)3/8} \approx 0.0267, A_{(2)1/2} \approx 0.3712$, $A_{(2)5/8} \approx 0.0141$, $A_{(2)3/4} \approx 0.0551$, $A_{(6)7/8} \approx 6.1775 \times 10^{-3}$, $A_{(2)1} \approx 0.1377$, $A_{(2)9/8} \approx 2.6234 \times 10^{-3}$, $A_{(2)5/4} \approx 0.0122$, $A_{(2)11/8} \approx 2.3976 \times 10^{-3}$, $A_{(2)3/2} \approx 0.0391$. The other amplitudes are $A_{(2)l/2} \in (10^{-15}, 10^{-2})$ ($l = 13, 14, \ldots, 128$) with $A_{(2)16} \approx 4.39 \times 10^{-14}$ except for $A_{(2)2} \approx 0.0110$. For such a period-4 evolution, one harmonic term can give an approximate solution. The analytical solution with 48 harmonic terms gives the period-8 evolutions with accuracy of $\varepsilon = 10^{-14}$. The main harmonic effects on the period-8 evolution are from $A_{(1)1/2} \approx 0.1551$, and $A_{(2)1/2} \approx 0.3712$.

4 Independent Periodic Evolutions

Consider a set of parameters as

$$a = 0.4, \ b = 1.2, \ Q_0 = 0.03. \tag{36}$$

Initial conditions for independent periodic evolutions are from the analytical solutions, as tabulated in Table 2.

The period-3 evolution with diffusion frequency $\Omega \approx 1.178$ are presented in Fig. 5. From the analytical solutions, the concentration orbits for stable and unstable period-3 evolutions are presented in Fig. 5i. One cycle is for period-3 evolution but is not a circle. The two concentration orbits for stable and unstable period-3 evolutions are similar. The harmonic spectrums for period-3 evolution of $\Omega \approx 1.178$ are presented in Fig. 6ii, iii. The constant for concentration x is $a_{(1)0/3} = 0.4$. The main

Fig. 4 Stable period-8
evolutions ($\Omega = 0.92$): (i)
concentration orbit (x, y), (ii)
harmonic spectrum for
concentration x, (iii)
harmonic spectrum for
concentration y. ($a =$
0.4, $b = 1.2$, $Q_0 = 0.08$),
IC: $(x_0, y_0) \approx$
(0.334620, 2.462824). The
solid curves and circular
symbols are for numerical
and analytical results

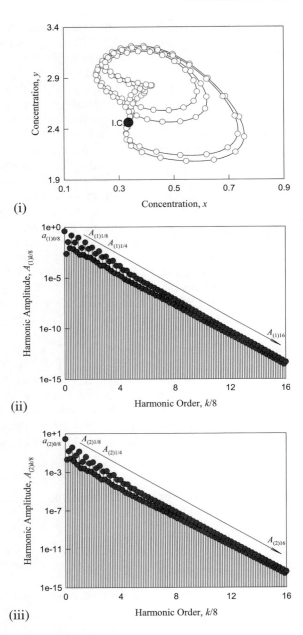

Table 2 Input data for numerical simulations ($a = 0.4$, $b = 1.2$, $Q_0 = 0.03$)

Ω	Initial conditions (x_0, y_0)	P-m evolution
1.178	(0.492590,2.796013)	P-3 (stable)
	(0.480758,2.866680)	P-3 (unstable)
0.977	(0.370809,3.214927)	P-5 (stable)
	(0.428270,3.113284)	P-5 (unstable)

Fig. 5 Independent period-3 evolutions: (i) Concentration orbit (x, y), (ii) harmonic amplitude for concentration x, (iii) harmonic amplitude for concentration y. ($a = 0.4$, $b = 1.2$, $Q_0 = 0.03$), IC: $(x_0, y_0) \approx$ (0.492590, 2.796013) (stable), $(x_0, y_0) \approx$ (0.480758, 2.866680) (unstable). Solid and dashed curves are for stable and unstable periodic evolutions

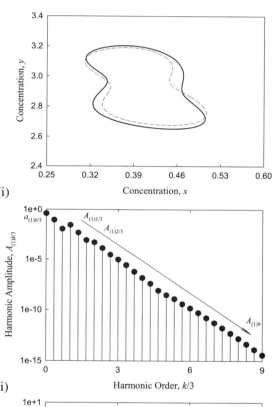

(i)

(ii)

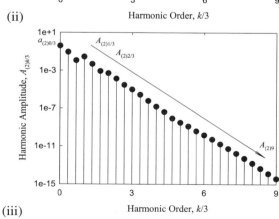

(iii)

Fig. 6 Independent period-5 evolutions: (i) concentration orbit (x, y), (ii) harmonic amplitude for concentration x, (iii) harmonic amplitude for concentration y. ($a = 0.4$, $b = 1.2$, $Q_0 = 0.03$). IC: $(x_0, y_0) \approx (0.370809, 3.214927)$ (stable), and $(x_0, y_0) \approx (0.428270, 3.113284)$ (unstable). Solid and dashed curves are for stable and unstable periodic evolutions

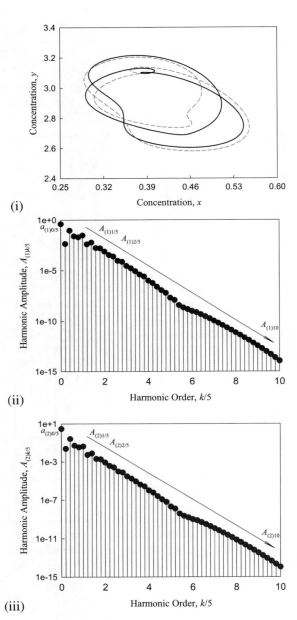

harmonic amplitudes are $A_{(1)1/3} \approx 0.0909$, $A_{(1)2/3} \approx 0.0116$, $A_{(1)3/3} \approx 0.0284$. The other harmonic amplitudes are $A_{(1)l/3} \in (10^{-14}, 10^{-3})$ for $l = 4, 5, \ldots, 27$ with $A_{(1)9} \approx 3.8 \times 10^{-15}$. The constant for concentration y is $a_{(2)0/5} \approx 2.9374$. The main harmonic amplitudes are $A_{(2)1/3} \approx 0.2487$, $A_{(2)2/3} \approx 0.0188$, $A_{(2)3/3} \approx 0.0251$. The other harmonic amplitudes are $A_{(2)l/3} \in (10^{-14}, 10^{-3})$ for $l = 4, 5, \ldots, 27$ with $A_{(2)9} \approx 3.9 \times 10^{-15}$. Such a periodic evolution at least needs 3 harmonic terms to get approximate solutions. The most effective harmonic terms are from $A_{(1)1/3} \approx 0.0909$ and $A_{(2)1/3} \approx 0.2487$. 27 harmonic terms are needed to keep the analytical period-3 evolutions are accurate.

The period-5 evolution with diffusion frequency $\Omega \approx 0.977$ are presented in Fig. 6 for illustration of oscillation complexity. From the analytical solutions, the concentration orbits for stable and unstable period-5 evolutions are presented in Fig. 6i. Solid and dashed curves are for stable and unstable periodic evolutions.

One cycle preserves the concave part on the left orbit while the right concave. One knot near the inner top orbit can be observed. The varying rate near such a knot is very slow, which implies that the period-5 evolution is a typical slow-fast varying evolution. The difference between the stable and unstable orbits can be clearly observed. In a similar fashion, the harmonic spectrums for period-5 evolution of $\Omega = 0.977$ are presented in Fig. 6ii, iii. The constant for concentration x is $a_{(1)0/5} = 0.4$. The main harmonic amplitudes are $A_{(1)1/5} \approx 4.3427 \times 10^{-3}$, $A_{(1)2/5} \approx 0.0911$, $A_{(1)3/5} \approx 0.0246$, $A_{(1)4/5} \approx 0.0198$, $A_{(1)5/5} \approx 0.0304$. The other harmonic amplitudes are $A_{(1)l/5} \in (10^{-14}, 10^{-3})$ for $l = 6, 7, \ldots, 50$ with $A_{(1)10} \approx 1.1 \times 10^{-14}$. The constant for concentration y is $a_{(2)0/5} \approx 2.9218$. The main harmonic amplitudes are $A_{(2)1/5} \approx 0.0226$, $A_{(2)2/5} \approx 0.2502$, $A_{(2)3/5} \approx 0.0487$, $A_{(2)4/5} \approx 0.0322$, $A_{(2)5/5} \approx 0.0399$. The other harmonic amplitudes are $A_{(2)l/5} \in (10^{-14}, 10^{-3})$ for $l = 6, 7, \ldots, 50$ with $A_{(2)10} \approx 1.1 \times 10^{-14}$. Such a periodic evolution at least needs 5 harmonic terms to get approximate solutions. The most effective harmonic terms are from $A_{(1)2/5} \approx 0.0911$ and $A_{(2)2/5} \approx 0.2502$.

5   Conclusions

The analytical solutions of periodic evolutions in the periodically forced Brusselator were obtained through the generalized harmonic balance method. Period-1 evolution to period-8 evolution on the bifurcation tree were presented, and the independent period-m evolutions ($m = 3, 5$) were presented for illustrations of complex periodic evolutions. Through the study, the method of generalized harmonic balance method is very good for nonlinear chemical oscillators, especially for slow-fast varying oscillators, which are difficult to be solved by the other analytical and numerical methods.

References

1. Belousov BP (1958) A periodic reaction and its mechanism. In: Collection of abstracts on radiation medicine. Medgiz, Moscow, pp 145–147
2. Zaikin AN, Zhabotinsky AM (1970) Concentration wave propagation in two-dimensional liquid-phase self-oscillating system. Nature 225(5232):535–537
3. Zhabotinsky AM (1964) Periodical oxidation of malonic acid in solution (a study of the Belousov reaction kinetics). Biofizika 9:306–311
4. Zhabotinsky AM (1964) Periodic liquid phase reactions. Proc Acad Sci USSR 157:392–395
5. Field RJ, Noyes RM (1972) Explanation of spatial band propagation in the Belousov reaction. Nature 237(5355):390–392
6. Tyson JJ (1973) Some further studies of nonlinear oscillations in chemical systems. J Chem Phys 58(9):3919–3930
7. Prigogine I, Lefever R (1968) Symmetry breaking instabilities in dissipative systems II. J Chem Phys 48(4):1695–1700
8. Lefever R, Nicolis G (1971) Chemical instabilities and sustained oscillations. J Theor Biol 30(2):267–284
9. Tomita K, Kai T, Hikami F (1977) Entrainment of a limit cycle by a periodic external excitation. Progress Theoret Phys 57(4):1159–1177
10. Nakhla M, Vlach J (1976) A piecewise harmonic balance technique for determination of periodic response of nonlinear systems. IEEE Trans Circ Syst 23(2):85–91
11. Kundert KS, Sangiovanni-Vincentelli A (1986) Simulation of nonlinear circuits in the frequency domain. IEEE Trans Comput Aided Des Integr Circuits Syst 5(4):521–535
12. Luo ACJ (2012) Continuous dynamical systems. Beijing/Glen Carbon, HEP/L&H Scientific
13. Luo ACJ, Huang JZ (2012) Analytical dynamics of period-m flows and chaos in nonlinear systems. Int J Bifurc Chaos 22(04):1250093
14. Luo ACJ, Huang JZ (2012) Approximate solutions of periodic motions in nonlinear systems via a generalized harmonic balance. J Vib Control 18(11):1661–1674
15. Luo ACJ, Huang JZ (2012) Analytical routines of period-1 motions to chaos in a periodically forced Duffing oscillator with twin-well potential. J Appl Nonlinear Dyn 1(1):73–108
16. Luo ACJ, Huang JZ (2012) Unstable and stable period-m motions in a twin-well potential Duffing oscillator. Discontin Nonlinearity Complex 1(2):113–145
17. Luo ACJ, Huang J (2013) Analytical period-3 motions to chaos in a hardening Duffing oscillator. Nonlinear Dyn 73(3):1905–1932
18. Luo ACJ, Lakeh AB (2013) Analytical solutions for period-m motions in a periodically forced van der Pol oscillator. Int J Dyn Control 1(2):99–115
19. Luo ACJ, Lakeh AB (2014) Period-m motions and bifurcation trees in a periodically forced, van der Pol-Duffing oscillator. Int J Dyn Control 2(4):474–493
20. Akhmet M, Fen MO (2017) Almost Periodicity in Chaos. Discontin Nonlinearity Complex 7(1):15–29
21. Luo ACJ, Guo SY (2018) Analytical solutions of period-1 to period-2 motions in a periodically diffused Brusselator. J Comput Nonlinear Dyn 13(9):090912

Analysis of Intermittent and Quasi-Periodic Transitions to Chaos in Vibro-Impact System with Continuous Wavelet Transform

Victor Bazhenov, Olga Pogorelova, and Tatiana Postnikova

Abstract The study of dynamical systems chaotic behavior and their routes to chaos was in particular attention at recent years. These phenomena study in non-smooth dynamical system is of the special scientists' interest. We apply the Continuous Wavelet Transform (hereinafter – CWT) to investigate the intermittent route to chaos, boundary crisis, and transitional regimes under quasi-periodic route to chaos in strongly nonlinear non-smooth discontinuous 2-DOF vibro-impact system. We show that CWT application is very useful to observe the intermittency, chaos and transitional regimes. It allows one to detect and to determine these phenomena with great confidence, reliability, and clearness.

1 Introduction

Chaotic dynamics is one of the most interesting and investigated subjects in nonlinear dynamics now. It is well known that just deterministic chaos is not an exceptional mode of dynamical systems behavior. On the contrary, chaotic behavior occurs in many dynamical systems in mathematics, mechanics, engineering, physics, chemistry, biology and medicine. Therefore, the chaotic dynamics study is one of the main ways of modern natural science development. Now, the theory of chaotic vibrations is well developed and is continuing to develop further [1–4].

It is a big honour for us to publish the chapter in the book devoted to the memory of Professor Valentin Afraimovich. He was studying the chaos and the routes to it. He was trying to understand the past and to predict the future.

V. Bazhenov · O. Pogorelova · T. Postnikova (✉)
Kyiv National University of Construction and Architecture, 31, Povitroflotskiy avenu, Kyiv, Ukraine
e-mail: posttan@ukr.net

V. Bazhenov
e-mail: vikabazh@ukr.net

O. Pogorelova
e-mail: pogos13@ukr.net

It is worth to underline that dynamical deterministic chaos occurs only when the control parameter changes in entirely deterministic systems and without any random external influence. This change may be very small.

Routes to chaos in nonlinear dynamical systems are of the special scientists' interest. Three main routes to chaos in dynamical systems are known [1, 3]:

- Period-doubling route to chaos – the most celebrated scenario, it is Feigenbaum scenario.
- Quasi-periodic route to chaos.
- Intermittent route to chaos by Pomeau and Manneville.

In addition, one can see that the chaotic attractor may face several critical stages, called crisis, where it undergoes sudden changes as a function of the parameter. At a crisis event the chaotic attractor is either vanishing or increasing in size. The phenomenon of crisis in chaotic attractors was first pointed out in [5]. Now three different types of crises are recognized [6]:

- Attractor merging crisis, where a multi-piece chaotic attractor merges together to increase size smoothly.
- Interior crisis, where a chaotic attractor increases in size abruptly.
- Boundary or exterior crisis, where a chaotic attractor suddenly vanishes.

The intermittency mechanisms, quasi-periodic routes to chaos, and crisis events were described in sufficient detail in the monographs [7, 8].

The dynamical behavior of two-body 2-DOF vibro-impact system (Fig. 1) was investigated during several years. Investigation results have been described in our previous papers [9–14]. This system has demonstrated intermittent and quasi-periodic routes to chaos, and boundary crisis. Continuous wavelet transform appeared very useful for its investigation.

Intermittency was discovered and divided into three types by French scientists Pomeau and Manneville [15]. When intermittency occurs, long regions of periodic motion with bursts of chaos are taking place. So zones of turbulent and laminar motion are alternating while the value of control parameter is the same. As one varies a control parameter, the chaotic bursts become more frequent and lasting. Intermittency classification is based on different types of local bifurcations where the periodic motion loses stability. Thus, intermittency type is defined by Floquet multiplier value. Depending on the point of the unit circle crossing, the type-I corresponds to crossing at $(+1)$; type-II – complex crossing; and type-III – crossing at (-1) [15].

Intermittency route to chaos has some complexity. First of all, it occurs less often than period doubling route. The latter occurs the most often and is studied the best. At second, "to catch" the intermittency in system motion is not a simple task. It is difficult to do it in a usual way. Intermittency shows broadband continuous Fourier spectrum, the set of Poincaré points, and the positive value of the largest Lyapunov exponent because of chaotic bursts in the large time range. Therefore, we apply the wavelet transform which copes with this task very well. It provides the time-frequency analysis or, more precisely, time-scale analysis of existing motion. CWT is useful to solve exactly this task.

While performing comparison for quasi-periodic and intermittent routes to chaos in our vibro-impact system, we see that the quasi-periodic route is considerably more intricate than the intermittent one. Different oscillatory regimes are replacing each other many times after Neimark-Sacker bifurcation while control parameter variations are very small. There are periodic subharmonic regimes – chatters, quasi-periodic, chaotic and transitional. What transitional regimes are? They occur in zones of transition from one regime to another and often correspond to prechaotic or postchaotic motion. They have inconsistent characteristics: their Fourier spectra fit to one regime type, Poincaré maps – to another. All of them have small value of the largest Lyapunov exponent. So we can't say something specified about their types.

Thus, we decided to test new technique in order to characterize this multiscale behavior. This new technique is Continuous Wavelet Transform (or CWT).

Wavelet analysis is useful to recognize periodic and chaotic motions in frequency and time domain. It turned out to be an efficient tool to study the vibro-impact system dynamic behavior. It provides the time-frequency signal representation and this is the main advantage. Therefore, it permits to detect unconformities and other abrupt signal changes [16–19].

The wavelet analysis has been applied in many scientific fields recent years. Some commercial software provides the wavelet analysis function, like Mathcad and Matlab [20].

We successfully applied CWT to study our vibro-impact system dynamic behavior. It was very useful to distinguish the periodic and the chaotic regimes. It helped us "to catch" the intermittency [21]. CWT gives the information about different frequencies' time dependence. So CWT copes with intermittency recognition well. This is a good reason to try to use CWT to study transitional regimes.

Chaotic motion and intermittency were studied with WT in different mechanical and physical systems in [22–30].

In [22], the authors apply the Discrete Wavelet Transform (or DWT) to the time series of variables for four hyper-chaotic systems. These are the Chen, Chua, Rössle and 5D systems. The presence of at least two positive Lyapunov exponents can confirm the existence of hyper-chaos. Hyper-chaotic systems are numerically simulated using the classical fourth-order Runge-Kutta algorithm. For wavelet analysis, the authors use the Daubeshies wavelet function db2. The authors regard the wavelet approach as a very illustrative means of revealing some dynamical properties of the hyper-chaotic systems. They believe that the information provided by the wavelet-scaling analysis can be used to choose the appropriate hyper-chaotic dynamic system for whatever application.

In an earlier paper [23], the authors used the DWT to obtain useful information about such chaotic systems as the Chua's system, Rössle system, and Chaotic Generator. We want to emphasize that the authors believe that "the method based on wavelet analysis is undoubtedly one of the most appealing, and successful ones". They hope that their results will inspire further use of wavelet analysis for chaotic time series.

In [24], the authors propose a method for detecting chaos in oscillatory circuits based on the wavelet transform. They introduce some specific measure obtained by averaging wavelet coefficients (AWC). This measure exhibits various values for chaotic and periodic states. The Chua's, Lorentz and Duffing chaotic oscillators are considered as examples. The Daubeshies wavelet order 12 is used for wavelet transform. The authors claim that the AWC-based algorithm for detecting current states (periodic or chaotic) allows one to detect chaos in a single time series, regardless of the system nature, and without a priori knowledge of the systems. The proposed algorithm accurately detects short intervals of chaotic or periodic behavior.

In [25], the authors used wavelet analysis to detect and localize unconformities events, or chaotic and periodic-cyclic sequences in stratigraphy. They apply the Continuous Wavelet Transform with the Morlet Wavelet. They emphasize that wavelet analysis is useful for recognizing a cyclic, chaotic, and monotonous process in both frequency and time domains. It has been proven to be an effective tool for cyclical stratigraphy. In particular, this makes it possible to detect unconformities and other abrupt changes in geological sequences.

In [26], the authors used wavelet analysis techniques based on a fifth-order biorthogonal Coiflets wavelet to study self-similarity and the intermittency of plasma fluctuations. They argue that wavelet-based techniques can be applied to all types of plasma fluctuation diagnostics.

There are several Russian papers on the application of wavelet transform for the intermittency study. In the works described by us, the authors use the Continuous Wavelet Transform to study the intermittency in nonlinear dynamic systems.

In [27, 28], type-I, II, III intermittency is studied using examples of discrete mappings. The effect of external noise on intermittency is investigated.

In [29], the mechanism for diagnosing the laminar and turbulent phases is illustrated by the traditional example, the Lorenz system. The authors believe that these results of wavelet analysis are sufficiently universal in the study of chaos.

In [30], the authors analyze the oscillations in nonlinear distributed systems – beams. They compare the use of the Morlet wavelet and the Gauss wavelet. It is shown that only wavelet transform with a Morlet wavelet gives accurate information about the nonlinear beams oscillations.

This brief review confirms the breadth and efficiency of the wavelet analysis use in various scientific fields. In particular, it is used to recognize periodic and chaotic oscillations, because its advantage lies in its simplicity of processing for studying irregular signals.

In this chapter we want to exhibit the following:

- to show that Continuous Wavelet Transform allows clearly distinguish the periodic and chaotic regimes of the vibro-impact system motion;
- to demonstrate the CWT ability to show the intermittency route to chaos with good visibility and reliability;
- to manifest transitional regimes at the quasi-periodic route to chaos.

2 The Short Description of Vibro-impact System Model

To achieve these goals we consider the model of strongly nonlinear non-smooth discontinuous 2-DOF two-body vibro-impact system (Fig. 1). Its dynamical behavior has been studied in our previous works [9–14]. The amplitude-frequency responses in wide range of control parameters [11] were obtained by parameter continuation method (Fig. 1). Here we'll give just short model description.

This model is formed by the main body m_1 and attached one m_2, which can play the role of percussive or non-percussive dynamic damper. Bodies are connected by linear elastic springs with stiffness k_1 and k_2 and dampers with damping coefficients c_1 and c_2. (The damping force is taken as proportional to first degree of velocity with coefficients c_1 and c_2).

The differential equations of its movement are:

$$\ddot{x}_1 = -2\xi_1\omega_1\dot{x}_1 - \omega_1^2 x_1 - 2\xi_2\omega_2\chi(\dot{x}_1 - \dot{x}_2) -$$
$$- \omega_2^2\chi(x_1 - x_2 + D) + \frac{1}{m_1}[F(t) - F_{con}(x_1 - x_2)], \qquad (1)$$
$$\ddot{x}_2 = -2\xi_2\omega_2(\dot{x}_2 - \dot{x}_1) - \omega_2^2(x_2 - x_1 - D) + \frac{1}{m_2}F_{con}(x_1 - x_2),$$

where $\omega_1 = \sqrt{\dfrac{k_1}{m_1}}$, $\omega_2 = \sqrt{\dfrac{k_2}{m_2}}$; $\xi_1 = \dfrac{c_1}{2m_1\omega_1}$, $\xi_2 = \dfrac{c_2}{2m_2\omega_2}$; $\chi = \dfrac{m_2}{m_1}$.

The main body is under effect of periodic external load $F(t) = P\cos(\omega t + \varphi_0)$, with period $T = \dfrac{2\pi}{\omega}$.

Impact is simulated by contact interaction force F_{con} according to contact quasistatic Hertz's law (2):

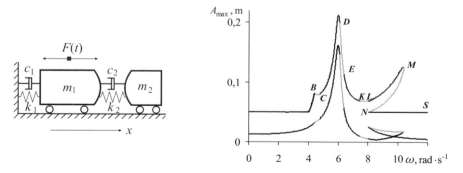

Fig. 1 Vibro-impact system model and amplitude-frequency responses

$$F_{con}(z) = K[H(z)z(t)]^{3/2},$$

$$K = \frac{4}{3}\frac{q}{(\delta_1 + \delta_2)\sqrt{A + B}}, \delta_1 = \frac{1 - v_1^2}{E_1\pi}, \delta_2 = \frac{1 - v_2^2}{E_2\pi}, \qquad (2)$$

where $z(t)$ is the bodies' closing in, $z(t) = x_2 - x_1$, A, B, and q – constants to characterize the local geometry of the contact zone; v_i and E_i – Poisson's ratios and Young's modulus for both bodies, $H(z)$ is the discontinuous Heaviside step function. Numerical parameters of this system are as following:

$$
\begin{array}{llll}
m_1 = 1000\,\text{kg}, & \omega_1 = 6.283\,\text{rad.s}^{-1}, & \xi_1 = 0.036, & E_1 = 2.1\cdot 10^{11}\,\text{N.m}^2, & v_1 = 0.3, \\
m_2 = 100\,\text{kg}, & \omega_2 = 5.646\,\text{rad.s}^{-1}, & \xi_2 = 0.036, & E_2 = 2.1\cdot 10^{11}\,\text{N.m}^{-2}, & v_2 = 0.3, \\
P = 500\,\text{N}, & A = B = 0.5\,\text{m}^{-1}, & q = 0.318. &
\end{array}
$$

The amplitude-frequency responses (Fig. 1) have some instability zones. The region DE demonstrates the intermittency route to chaos and boundary crisis, the quasi-periodic transition to chaos occurs in section KL.

3   Intermittent Route to Chaos

Points D and E of the amplitude-frequency responses (Fig. 1) – are those where the main periodic (1,1)-regime is losing stability (regime with period $1T$ and 1 impact per cycle). Floquet multiplier μ_1 is real, it crosses the unit circle along the real axis at point (-1). The intermittency that will take place will be the type-III intermittency.

There are two plots showing the whole motion picture in DE region with good visibility (Figs. 2 and 3).

Figures show how system dynamic states are changing with the control parameter. Let us have an attentive look at the largest Lyapunov exponent λ dependence on control parameter, which is the external loading frequency ω (Fig. 2). It is well known that Lyapunov exponents are used to characterize the type of dynamical system motion. In order to have a chaos criterion, one needs to calculate the largest exponent only.

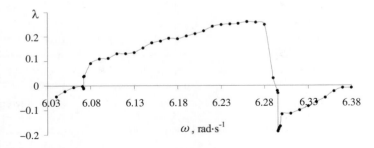

Fig. 2 The largest Lyapunov exponent dependence on control parameter

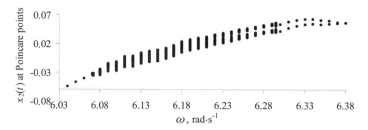

Fig. 3 Bifurcation diagram

In case of $\lambda > 0$ the nearby phase trajectories diverge. If $\lambda < 0$ they converge on the average. The λ sign is the chaos criterion. For regular motions $\lambda \leq 0$, but for chaotic motion $\lambda > 0$. The positive value of the largest Lyapunov exponent imply the chaotic dynamics. We reported on the largest Lyapunov exponent estimation in non-smooth vibro-impact system in [13]. Figure 2 shows that the positive value of the largest Lyapunov exponent corresponds to the motion influenced by the external frequency ranged $6.07 \, \text{rad} \cdot \text{s}^{-1} \leq \omega \leq 6.29 \, \text{rad} \cdot \text{s}^{-1}$. Is it the full chaos in whole frequency range? And how the transition to chaos is going on? We'll give answers later.

Figure 3 shows the bifurcation diagram. This is a widely used technique to investigate different states of a dynamical system with a parameter variation. As one can see in Fig. 3 the control parameter value is plotted on the horizontal axis, and Poincaré points phase coordinates $x_2(t)$ are plotted on the vertical axis. There is only one value of one point coordinate in Poincaré map for (1,1)-regime. We see one point along vertical line at bifurcation diagram for $\omega < 6.07 \, \text{rad} \cdot \text{s}^{-1}$ and $\omega \geq 6.38 \, \text{rad} \cdot \text{s}^{-1}$. There are 2 separate points along vertical line for periodic regimes (2,2) and (2,3). These are regimes with $2T$ period and 2 or 3 impacts per cycle. One can see solid vertical lines for chaotic regimes.

Under $\omega = 6.07 \, \text{rad} \cdot \text{s}^{-1}$ and $\omega = 6.29 \, \text{rad} \cdot \text{s}^{-1}$ the (1,1)-regime becomes (2,2)-regime. Under $\omega = 6.30 \, \text{rad} \cdot \text{s}^{-1}$ it becomes (2,3)-regime. We don't observe the further period doubling. Feigenbaum's route to chaos doesn't realize under these frequencies. But then we observe type-III intermittency under some frequencies inside the allegedly chaotic motion.

As one can see at the left border of the Figs. 2 and 3, when $\omega = 6.07 \, \text{rad} \cdot \text{s}^{-1}$ the periodic (1,1)-regime loses stability, real Floquet multiplier is equal (-1). The (2,2)-regime arises when the control parameter ω is increasing. Then intermittency appears simultaneously with period doubling. The amplitude of subharmonic is growing, and the amplitude of main harmonic is decreasing (Table 1) [4]. When the subharmonic amplitude becomes big, signal loses regularity, and turbulent bursts arise.

At the beginning of intermittency narrow and rare turbulent bursts occur in between of wide laminar phases. They are the bursts of vibrations with low frequencies and small amplitudes. These bursts have pale color at the plots of the wavelet surface projections. It is type-III intermittency. The wavelet surface projections in wide and narrow time ranges demonstrate this phenomenon very well (Fig. 4).

Table 1 Main harmonic and subharmonic amplitudes before intermittency

ω (rad·s^{-1})	Amplitude (m)	$\omega/2$ (rad·s^{-1})	Amplitude (m)	Regime
6.0700	0.1560	3.03500	0.0879	(2.2)
6.0710	0.1542	3.03550	0.1018	(2.2)
6.0715	0.1533	3.03575	0.1076	(2.2)
6.0716	0.1532	3.03580	0.1082	(2.2)
6.0717	0.15340	3.03585	0.1065	Intermittency
6.0718	0.15337	3.03590	0.1067	Intermittency
6.0719	0.15363	3.03595	0.1046	Intermittency
6.0720	0.15355	3.03600	0.1051	Intermittency

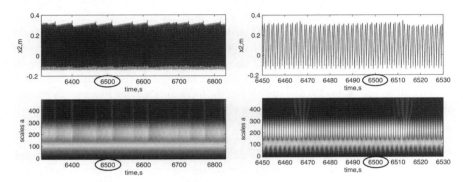

Fig. 4 Time histories and wavelet surface projections for intermittency under $\omega = 6.076$ rad·s^{-1}, $\lambda = 0.039$ (Color while on screen)

It is worth to stress once more that CWT turned out very useful and convenient tool for intermittency discovery and visualization. These plots give us a clear picture of frequencies distribution in time.

We use wavelet Morlet for Continuous Wavelet Transform. Here and further all the plots are given for attached body. Its mass is much lower than the main body mass. So oscillatory amplitudes are bigger, and changes are better seen, thus, the plots are more obvious ones.

With the ω increase, the turbulence grows quickly, the turbulent bursts become more frequent. Subharmonic disappears sometimes, and only vibrations with main frequency remain in laminar phases.

We show the intermittency under $\omega = 6.13$ rad·s^{-1} where it is well-pronounced (Fig. 5).

Oscillogram in Fig. 5 is typical for type-III intermittency [4]. Let us have an attentive look at wavelet surface projection. Zones where chaotic motion is in turbulent phases are well observed. High and low frequencies are interrupted; only one high frequency remains. In other words, laminar regions occur interrupted by chaos. Small

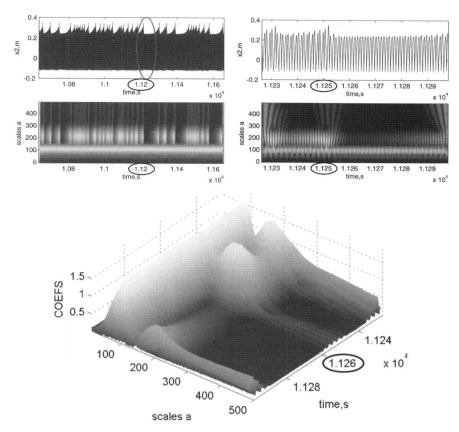

Fig. 5 Time histories, wavelet surface projections, and surface of wavelet coefficients inside the red oval for intermittency under $\omega = 6.13\,\text{rad·s}^{-1}$, $\lambda = 0.126$ (Color while on screen)

time interval picked out by red oval is also shown in Fig. 5. Here we also see the sharp transformation of chaotic motion into almost periodic one.

It is worth to underline once more that it is difficult to "catch" the intermittency in usual way. Intermittency shows broad continuous Fourier spectrum, the set of Poincaré points, and the positive value of the largest Lyapunov exponent because of chaotic bursts in the large time range. We see that not the whole frequency range $6.07\,\text{rad·s}^{-1} \le \omega \le 6.29\,\text{rad·s}^{-1}$ corresponds to the full chaos in spite of the Lyapunov exponent positive sign (Fig. 2). There are regions of intermittency. But the wavelet transform copes with the intermittency recognition task very well because it gives the frequency-time picture of the existing motion.

Now, let us have a more attentive look to the regimes during intermittency. In Fig. 6 we show the phase trajectories, Poincaré maps, Fourier spectra in logarithmic scale, and the surfaces of wavelet coefficients for two different regimes. Both – chaotic and the periodic – exists inside the red oval.

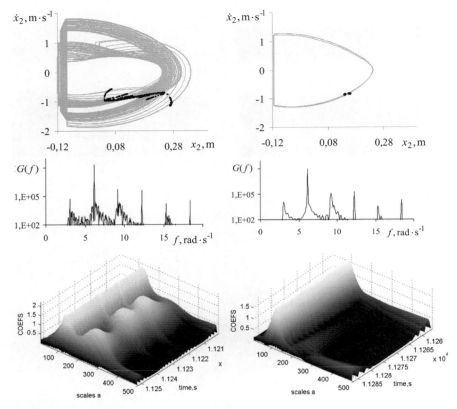

Fig. 6 Phase trajectories with Poincaré maps, Fourier spectra, and surfaces of wavelet coefficients for chaotic and periodic regimes inside the red oval under intermittency ($\omega = 6.13\,\text{rad·s}^{-1}$, $\lambda = 0.126$) (Color while on screen)

Then the full chaos begins. Time histories, wavelet surface projection, and surface of wavelet coefficients for chaotic motion under $\omega = 6.20$ rad·s^{-1} are shown in Fig. 7.

Please note that numbers on the scale axis for these and further plots of wavelet characteristics are not corresponding rad·s^{-1}. They are just the numbers of samples given by the computation in the MATLAB program CWT.

We see a lot of vibrations with low frequency and weak power which form the continuous Fourier spectrum. They are not constant but are changing with the time. The higher frequency (subharmonic) is not constant with the time too. It is typical for non-regular motion. To confirm this motion chaoticity we show its phase trajectories, Poincaré map and Fourier spectrum in logarithmic scale in Fig. 7.

Thus, we see that wavelet transform works really well to discern different regimes.

Now, let's have a look at the right border of bifurcation diagram and Lyapunov exponent plot in Figs. 2 and 3. Under $\omega = 6.28$ rad·s^{-1} the chaotic attractor suddenly vanishes. We almost immediately see the periodic (2,3)-regime, and then the

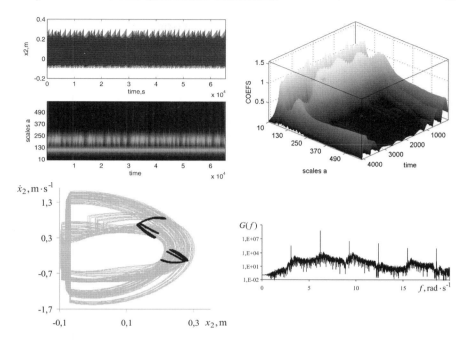

Fig. 7 Time histories, wavelet surface projection, surface of wavelet coefficients, phase trajectories with Poincaré map, and Fourier spectrum for chaotic motion under $\omega = 6.20\,\mathrm{rad\cdot s^{-1}}$, $\lambda = 0.213$ (Color while on screen)

periodic (2,2)-regime in a narrow frequency range. Under $\omega = 6.37\,\mathrm{rad\cdot s^{-1}}$ Floquet multiplier is back into unit circle crossing it at point (-1). The stable periodic (1, 1)-regime arise. It is worth to point out that only when frequency change is very small ($\omega = 6.295$, $6.290\,\mathrm{rad\cdot s^{-1}}$) we see transitional regimes similar to the chatter: many points in Poincaré map, the largest Lyapunov exponent with a small value, but the Fourier spectrum is broad and continuous, the number of low frequencies at wavelet characteristics is big. We'll describe the transitional regimes in detail in Sect. 4. The scheme of regimes replacement is presented in Table 2.

So we consider such transition to chaos as boundary (or exterior) crisis, where a chaotic attractor suddenly vanishes [6, 8].

Thus, we see that Continuous Wavelet Transform copes very well with the intermittency detection; it's a very useful and convenient tool to observe the intermittency and chaos in vibro-impact system. It gives the possibility to detect and to identify these phenomena with great confidence and reliability. It obviously demonstrates the intermittent route to chaos, allows to distinguish and to analyze the laminar and turbulent phases. Plots of wavelet coefficients surfaces and their projections present these regimes brightly, especially the colour plots while on screen. The wavelet theory and existing Software are very useful to study these phenomena.

Table 2 The scheme of regimes replacement under boundary crisis

ω (rad·s^{-1})	6.280	6.290	6.295	6.297	6.300	6.310	6.330	6.370	6.380
Regime	Chaotic	Transitional (chatter?)	Transitional (chatter?)	(2,3)	(2,3)	(2,2)	(2,2)	(2,2)	(1,1)
λ	0.249	0.031	−0.022	−0.173	−0.115	−0.114	−0.085	−0.01	−0.009

4 Transitional Regimes Under Quasi-Periodic Route to Chaos

Stable periodic (1,1)-regime loses its stability at points K and L of the amplitude-frequency responses (Fig. 1). Quasi-periodic regimes arise as a result of Neimark-Sacker bifurcations. Two complex conjugate multipliers μ and μ^* leave the unit circle [12, 13].

Talking about quasi-periodic route to chaos one has to keep in mind that the whole picture in this case becomes essentially complicated. Many aspects are still not studied in a full. Attractor evolution with the governing parameter change may be intricate. Quasi-periodic and periodic regimes may alternate and undergo different bifurcations. Vibro-impact system exhibits a rich variety of dynamics.

We got the largest Lyapunov exponent dependence on control parameter where excitation frequency ω is used as control parameter (Fig. 8).

This plot demonstrates the quasi-periodic route to chaos very obviously. It is more intricate and manifold than intermittent route to chaos (Fig. 2). After the torus breakup, different oscillatory regimes replace each other many times under very small control parameter variation. There are periodic regimes with long periods and a lot of impacts per cycle (chatters) – (19, 12), (9, 6), (14, 10), and (23, 17)-regimes. There are two zones of hysteresis, and two zones of transitional regimes. We observed chaotic motions in narrow frequency ranges.

Quasi-periodic route to chaos in our vibro-impact system was described in detail in [12, 13].

The birth and breakdown of an invariant torus was studied in works such as [31–35].

In [32], the author considers a simple system of two nonlinearly coupled oscillators using the techniques of averaging and numerical bifurcation. The dynamics of what has occurred after the Neimark-Sacker bifurcation is studying. The author shows how the emerging torus becomes unsmooth before it breaks down and numerically shows how the torus gets destroyed. The author writes that the torus destruction occurs according to one of the Afraimovich-Shilnikov scenarios [31].

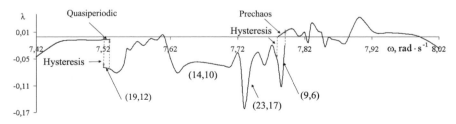

Fig. 8 The largest Lyapunov exponent dependence on control parameter

In [33], the author describes how bifurcations of periodic solutions generate a torus, and transitions like the Neimark-Sacker bifurcations, starting from a periodic solution, generate quasi-periodic solutions.

The breakdown of a torus starts with its non-smoothness. A number of examples illustrate the theory.

Authors of the well-known paper [31] consider three possible ways of torus breakdown. At the end they conclude: "The further experimental and numerical investigations have to confirm or refute our hypotheses. "These possible ways were interpreted in [33] as following: "...the breakdown of the torus starts with non-smoothness of the torus. The authors [31] give three possibilities:

• Stable and unstable periodic orbits vanish through a bifurcation.
• Stable and unstable manifolds of the unstable periodic orbit intersect tangentially to form a homoclinic orbit.
• Stable periodic orbit loses stability."

Our numerical investigations of torus breakdown [12, 13] give us the reason to believe that two positions can be applied to our results:

• Breakdown of the torus starts with its non-smoothness.
• Stable periodic orbit loses stability.

There is the chapter "Universal quasi-periodic route to chaos" in the well-known book [4]. Although our numerical experiment is devoted to this topic, we failed unfortunately to find a straight correlation between our results and Shuster's conclusions.

We also think that conclusion from [33] may be applied to our results too: "An interesting side-remark is that the phenomena described here, confirm a visionary paper by Ruelle and Takens [36] who suggested a new bifurcation scenario where a periodic solution produces subsequently a torus and then a strange attractor."

Now, we'll discuss the transitional regimes under this route to chaos. What the transitional regimes are? They occur in transition zones from one regime to another, and often correspond to prechaotic or postchaotic motion. They have inconsistent characteristics: their Fourier spectra fit one regime type, Poincaré maps – another one. Value of their largest Lyapunov exponent is small. Since Continuous Wavelet Transform appeared very useful to detect the intermittency and chaotic motion, it is logical to try to apply it to transitional regimes recognition with bigger precision.

At first let's consider the left zone of transitional regimes where ω is around $7.61\,\text{rad}\cdot\text{s}^{-1}$. Figure 9 shows the phase trajectories with Poincaré map and Fourier spectrum in logarithmic scale (for attached body m_2) for transitional regime under $\omega = 7.61\,\text{rad}\cdot\text{s}^{-1}$. Its Poincaré map is almost closed curve – as if the regime is quasi-periodic. But the Fourier spectrum is broadband and continuous, as if the motion is chaotic. These characteristics are inconsistent. This regime is nor quasi-periodic, neither chaotic one. Can wavelet characteristics depict its type with bigger precision? The wavelet surface projection and surface of wavelet coefficients for this regime are also shown in Fig. 9. We see two high frequencies with high power and a lot of low frequencies with very weak power which provide the continuous Fourier spectrum.

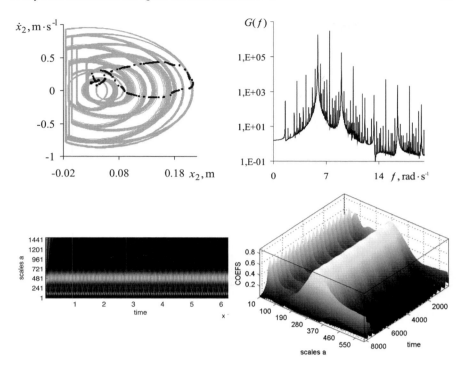

Fig. 9 Phase trajectories with Poincaré map, Fourier spectrum, wavelet surface projection, and surface of wavelet coefficients for transitional regime under $\omega = 7.61 \, \text{rad·s}^{-1}$, $\lambda = 0.0027$ (Color while on screen)

They change a little with the time that provides almost closed curve on Poincaré map. These plots don't give an idea to clarify this regime type. But they confirm that this regime is neither quasi-periodic nor chaotic one. It is exactly transitional regime.

Now, let's investigate the second region of transitional motions where $7.815 \, \text{rad·s}^{-1} < \omega < 7.90 \, \text{rad·s}^{-1}$. Different but very similar modes alternate in this frequency range while the control parameter is changing a little. We see chaotic regimes for $\omega = 7.80, 7.81, 7.815 \, \text{rad·s}^{-1}$. Characteristics for such regime under $\omega = 7.815 \, \text{rad·s}^{-1}$ are shown in Fig. 10.

These plots look typical for chaotic motion. Poincaré map is a disordered set of points in the limited space. Wavelet characteristics plots show two high frequencies with high power and a lot of low frequencies with weak power that provide the continuous Fourier spectrum. It is seen well that all of them are changing with the time.

Then immediately we see transitional (prechaotic) regime under $\omega = 7.82$, $7.825 \, \text{rad·s}^{-1}$. Its Poincaré map has the set of separate points, but its Fourier spectrum is a continuous one.Maybe we can say that this is "chatter"? Its characteristics under $\omega = 7.82 \, \text{rad·s}^{-1}$ are depicted at Fig. 11.

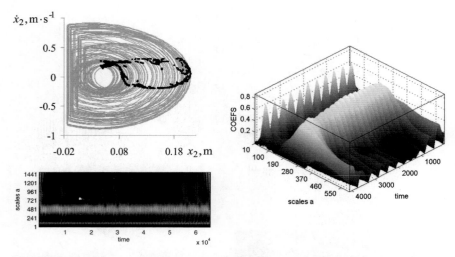

Fig. 10 Phase trajectories with Poincaré map, wavelet surface projection, and surface of wavelet coefficients for chaotic regime under $\omega = 7.815\,\text{rad}\cdot\text{s}^{-1}$, $\lambda = 0.0031$ (Color while on screen)

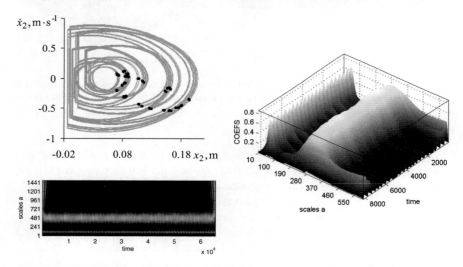

Fig. 11 Phase trajectories with Poincaré map, wavelet surface projection, and surface of wavelet coefficients for transitional regime under $\omega = 7.82\,\text{rad}\cdot\text{s}^{-1}$, $\lambda = 0.0092$ (Color while on screen)

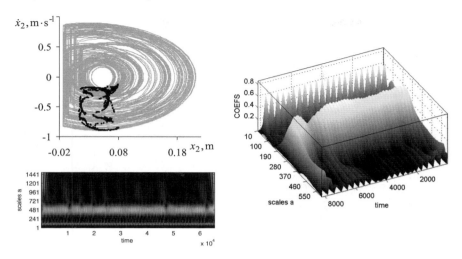

Fig. 12 Phase trajectories with Poincaré map, wavelet surface projection, and surface of wavelet coefficients for chaotic regime under $\omega = 7.83\,\text{rad·s}^{-1}$, $\lambda = 0.031$ (Color while on screen)

Modes alternate in this frequency range very quickly. Only with one frequency value $\omega = 7.83\,\text{rad·s}^{-1}$ immediately after the transitional regime, we again see chaotic motion (Fig. 12).

Wavelet plots for these regimes differ little from each other. Only the Poincaré maps are noticeably different. These regimes are similar to each other. They have two high frequencies with high power and many low frequencies with weak power that provide the broad continuous Fourier spectrum.

Immediately after the chaotic regime, the transitional ones with sets of separate points on the Poincaré maps and with continuous Fourier spectra exist when $\omega = 7.84$, 7.845, $7.85\,\text{rad·s}^{-1}$, and further to $\omega = 7.920\,\text{rad·s}^{-1}$. They are prechaotic or postchaotic regimes. The characteristics of such regime with $\omega = 7.845\,\text{rad·s}^{-1}$ are shown in Fig. 13.

Only further in a narrow frequency range of $7.90\,\text{rad·s}^{-1} \leq \omega \leq 7.92\,\text{rad·s}^{-1}$ does a truly chaotic motion occur. We considered this in detail in [10, 11], so now we will not discuss it. We only note the following. It has an excellent Poincaré map, which we called "Leaflet de Poincaré", by analogy with "Fleur de Poincaré", by Professor Francis Moon [1]. This Poincaré map demonstrates the fractal structure [3, 37]. At least, this structure is very similar to fractal. And its wavelet characteristics are typical for chaotic motion (Fig. 14).

We see that all these transitional regimes are similar to each other. They have small value of the largest Lyapunov exponent, the set of separate points or the almost closed curve on the Poincaré map, and the broad continuous Fourier spectrum. The characteristics obtained with the Continuous Wavelet Transform are also very similar. We see two high frequencies with high power and many low frequencies with very weak power, which provide a continuous Fourier spectrum. They change little in

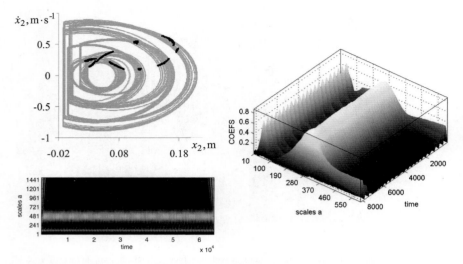

Fig. 13 Phase trajectories with Poincaré map, wavelet surface projection, and surface of wavelet coefficients for transitional regime under $\omega = 7.845\,\text{rad·s}^{-1}$, $\lambda = 0.0086$ (Color while on screen)

time, and this provides an almost closed curve or a set of separate points on the Poincaré map.

5 Conclusions

- The continuous wavelet transform is very useful to observe the intermittency, boundary crisis, and chaos in vibro-impact system. It makes possible to detect and to identify these phenomena with great confidence and reliability. It clearly demonstrates the intermittent route to chaos, allows one to distinguish and to analyze laminar and turbulent phases. Wavelet coefficient surfaces and their projection plots give a very clear presentation of these regimes, especially the colour plots on screen. The wavelet theory and existing Software are very useful to study these phenomena.
- Transitional regimes are similar. They have small value of the largest Lyapunov exponent, almost closed curve or system of separate points on Poincaré map, and the broad continuous Fourier spectrum. The characteristics obtained with Continuous Wavelet Transform are very much alike. CWT plots confirm these regimes nor quasi-periodic neither chaotic one. They are exactly transitional regimes. Sometimes they are chatter alike but have a lot of low frequencies with very weak power. CWT plots clearly demonstrate different frequencies presence in time histories and their time distribution. Thus, Continuous Wavelet Transform is a very effective tool to recognize what one or another transitional regime is.

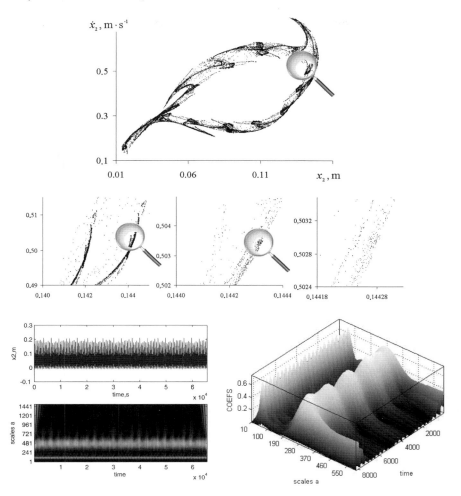

Fig. 14 Poincaré map, fractal structure, time histories, wavelet surface projection, and surface of wavelet coefficients for transient chaos ($\omega = 7.92\,\mathrm{rad\cdot s^{-1}}$, $\lambda = 0.014$) (Color while on screen)

References

1. Moon FC (1987) Chaotic vibrations: an introduction for applied scientists and engineers. Wiley, New York
2. Thompson JMT, Stewart HB (2002) Nonlinear dynamics and chaos. Wiley
3. Kuznetsov SP (2006) Dynamical chaos. Fizmatlit, Moscow
4. Schuster HG (1988) Deterministic chaos: an introduction. Physik-Verlag GmbH
5. Grebogi C, Ott E, Yorke JA (1983) Crises, sudden changes in chaotic attractors, and transient chaos. Physica D Nonlinear Phenomena 7(1–3):181–200
6. Ananthkrishnan N, Sahai T (2001) Crises-critical junctures in the life of a chaotic attractor. Resonance 6(3):19–33

7. Nayfeh Ali H, Balakumar B (2008) Applied nonlinear dynamics: analytical, computational, and experimental methods. Wiley
8. Hilborn RC (2000) Chaos and nonlinear dynamics: an introduction for scientists and engineers. Oxford University Press on Demand
9. Bazhenov VA, Pogorelova OS, Postnikova TG (2017) Stability and discontinious bifurcations in vibroimpact system: numerical investigations. LAP LAMBERT Academic Publishing GmbH and Co. KG Dudweiler, Germany
10. Bazhenov VA, Lizunov PP, Pogorelova OS, Postnikova TG, Otrashevskaia VV (2015) Stability and bifurcations analysis for 2-DOF vibroimpact system by parameter continuation method. Part I: loading curve. J Appl Nonlinear Dyn 4(4):357–370
11. Bazhenov VA, Lizunov PP, Pogorelova OS, Postnikova TG (2016) Numerical bifurcation analysis of discontinuous 2-DOF vibroimpact system. Part 2: frequency-amplitude response. J Appl Nonlinear Dyn 5(3):269–281
12. Bazhenov VA, Pogorelova OS, Postnikova TG (2019) Breakup of closed curve—quasi-periodic route to chaos in vibroimpact system. Discontin Nonlinearity Complex 8(3):275–288
13. Bazhenov VA, Pogorelova OS, Postnikova TG (2019) Quasi-periodic route to transient chaos in vibroimpact system. In: Luo ACJ, Volchenkov D (eds) Nonlinear dynamics, chaos, and complexity I: in memory of Valentin Afraimovich (1945–2018). Higher Education Press
14. Bazhenov VA, Pogorelova OS, Postnikova TG (2019) Intermittent and quasi-periodic routes to chaos in vibroimpact system: numerical simulations. LAP LAMBERT Academic Publishing GmbH and Co. KG Dudweiler, Germany
15. Pomeau Y, Manneville P (1980) Intermittent transition to turbulence in dissipative dynamical systems. Commun Math Phys 74(2):189–197
16. Polikar R (1996) The wavelet tutorial. Ames, Jowa
17. Chui CK (2016) An introduction to wavelets. Elsevier
18. Daubechies I (1992) Ten lectures on wavelets. In: CBMS-NSF series in applied mathematics 61. SIAM, Philadelphia
19. Astafieva NM (1996) Wavelet analysis: basic theory and some applications. Uspekhi fizicheskikh nauk 166(11):1145–1170
20. https://www.mathworks.com/help/wavelet/ref/cwt.html
21. Bazhenov VA, Pogorelova OS, Postnikova TG (2018) Intermittent transition to chaos in vibroimpact system. J Appl Math Nonlinear Sci 3(2):475–486
22. Murguía JS, Campos-Cantón E et al (2006) Wavelet analysis of chaotic time series. Revista Mexicana de Física 52(2):155–162
23. Murguía JS et al (2018) Wavelet characterization of hyper-chaotic time series. Revista Mexicana de Física 64(3):283–290
24. Rubežić V, Djurović I, Sejdić E (2017) Average wavelet coefficient-based detection of chaos in oscillatory circuits. COMPEL Int J Comput Math Electr Electr Eng 36(1):188–201
25. Prokoph A, Barthelmes F (1996) Detection of nonstationarities in geological time series: wavelet transform of chaotic and cyclic sequences. Comput Geosci 22(10):1097–1108
26. Xu GS, Wan BN, Zhang W (2006) Application of wavelet multiresolution analysis to the study of self-similarity and intermittency of plasma turbulence. Rev Sci Instrum 77(8):083505
27. Afonin VV, Boletskaya TK (2011) Exploitation of wavelet analysis for investigation of intermittency in dynamical nonlinear systems. Vestnik Novosibirsk State University, Physics 6(2):93–97 (in Russian)
28. Afonin VV, Boletskaya TK (2011) Wavelet-analysis of II and III type intermittency. Nonlinear Dyn 7(3):427–436 (in Russian)
29. Koronovskii AA, Khramov AE (2001) An effective wavelet analysis of the transition to chaos via intermittency. Tech Phys Lett 27(1):1–5 (in Russian)
30. Krysko AV, Zhigalov MV, Soldatov VV, Podturkin MN (2009) The best wavelet selection at the nonlinear flexible beams vibrations analysis with transversal displacement. Bull Saratov State Tech Univ 3(1)
31. Afraimovich VS, Shilnikov LP (1991) Invariant two-dimensional tori, their breakdown and stochasticity. Amer Math Soc Transl 149(2):201–212

32. Bakri, T (2005) Torus breakdown and chaos in a system of coupled oscillators. Int J Non-Linear Mech
33. Verhulst F (2016) Torus break-down and bifurcations in coupled oscillators. Procedia IUTAM 19:5–10
34. Bakri T, Kuznetsov YA, Verhulst F (2015) Torus bifurcations in a mechanical system. J Dyn Diff Equ 27(3–4):371–403
35. Bakri T, Verhulst F (2014) Bifurcations of quasi-periodic dynamics: torus breakdown. Zeitschrift für angewandte Mathematik und Physik 65(6):1053–1076. Springer
36. Ruelle D, Takens F (1971) On the nature of turbulence. Les rencontres physiciens-mathématiciens de Strasbourg-RCP25 12:1–44
37. Feder J (2013) Fractals. Springer Science & Business Media

Complex Dynamics of Solitons in Rotating Fluids

Lev A. Ostrovsky and Yury A. Stepanyants

Abstract Interaction of solitons with a long background wave is studied within the framework of rotation modified Korteweg–de Vries (rKdV) equation. Using the asymptotic method for solitons interacting as classical particles with a long wave, we derive a set of ODEs describing soliton amplitude and its phase with respect to the background wave. The shape of the background wave may range from a sinusoidal to the limiting profile representing a periodic sequence of parabolic arcs, which is specifically considered here. We analyze energy exchange between a soliton and the long wave taking into account the radiation losses of solitons. It is shown that the losses can be compensated by energy pumping from the long wave and, as the result, a stationary soliton can exist, unlike the case when there is no spatially variable background. Then the complex dynamics of two solitons on a long wave is analyzed for two basic cases: solitons with strongly different amplitudes (overlapping interaction) and solitons with close amplitudes (exchange interaction), including their behavior in time, phase space, and Fourier spectra of their amplitudes.

1 Introduction

Currently it is a well-known fact that solitary waves (or solitons) reveal the features similar to those of material particles [4, 14–16]. In particular, two interacting solitons in the integrable systems emerge from their collision with the same parameters as before the collision, similarly to the elastic collision of two classical particles [26], which led to the term "soliton". We believe that the same notion can be extended to solitary waves in non-integrable systems; although their collisions are not entirely

L. A. Ostrovsky (✉)
University of Colorado, Boulder, USA

Institute of Applied Physics of the Russian Academy of Sciences, Nizhny Novgorod, Russia

Y. A. Stepanyants
School of Sciences, Faculty of Health, Engineering and Sciences, University of Southern Queensland, Toowoomba, QLD 4350, Australia
e-mail: Yury.Stepanyants@usq.edu.au

"elastic" since they can radiate small-amplitude waves. For the integrable equations, exact multi-soliton solutions can often be obtained by the inverse scattering method and some other techniques such as the Hirota transform, Bäcklund transform, or Darboux–Matveev transform [1, 19].

Along with the exact methods, the approximate, asymptotic methods have been developed for similar problems, which are equally applicable to the integrable and non-integrable systems where solitary waves behave as the attractive or repulsive particles, and sometime demonstrating even more complex formations, such as stationary or oscillating soliton pairs, multi-soliton structure, and breathers [4, 6–8, 14–16].

A radically different dynamics of impulses close to solitons occurs in rotating systems such as surface and internal gravity waves in the ocean under the action of the Coriolis force. As demonstrated by field data [3, 13, 18] and laboratory experiments [10, 12, 24], the effects of nonlinearity and rotation-induced dispersion can be comparable in many real situations. Modeling of such waves leads to the apparently non-integrable rotation modified Korteweg–de Vries (rKdV) equation [20] which, strictly speaking, does not allow solitary waves at all due to radiation losses. However, solitons can exist on a long-wave background, which pumps up the energy into a soliton and thus compensates the losses. In such a configuration, a soliton can perform complex motions, both periodic and unlimited. Even more complex is the dynamics of two or more solitons superimposed on a long wave. Some interesting aspects of this dynamics are briefly discussed below.

2   The Background

2.1   *Soliton Interactions Without Rotation*

First we give some basic information about the behavior of solitons as particles. As mentioned, for the Korteweg–de Vries (KdV) and other integrable equations, interaction of solitons can be described by exact methods. The asymptotic theory of such processes is most thoroughly developed for the cases when the well separated solitons interact via their asymptotics ("tails") [4, 6–8]. In these cases, the energy always flows from the rear soliton to the frontal one, and two scenarios are possible: (1) Overtaking, when the rear soliton is strong enough to overtake the frontal one, and (2) exchange, when the frontal soliton is accelerated enough to break away. These processes are schematically shown in Fig. 1. The theory described below includes this process as a particular case.

As mentioned, in the integrable systems, the solitons eventually restore the same amplitudes as before the interaction so that the latter only causes an additional time delay or phase shift (the "elastic collision"). In non-integrable systems such processes are commonly accompanied by radiation of small-amplitude wave trains; therefore,

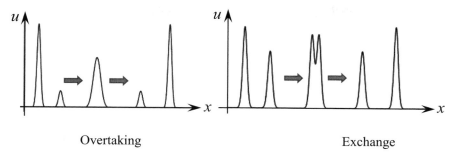

Overtaking   Exchange

Fig. 1 Two types of interaction of two solitons

soliton parameters after the interaction slightly differ from those before (the "inelastic collision").

2.2   Rotating Media

Earth rotation makes the problem much more complex. In what follows we base our consideration on the rKdV model equation obtained as early as in 1978 [20] and now becoming an object of numerous studies:

$$(u_t + c_0 u_x + \alpha u u_x + \beta u_{xxx})_x = \gamma u, \tag{2.1}$$

where c_0 is a linear long-wave velocity, α and β are, respectively, the coefficients of nonlinearity and dispersion, and γ is proportional to the squared Coriolis frequency.

This equation is not known to be integrable (except for the limits of $\gamma = 0$ or $\beta = 0$); its solutions are defined by the interplay of "two dispersions": the Boussinesq-type ($\sim\beta$) and Coriolis-type ($\sim\gamma$). Some specific features of this model found in different years are:

1. For the periodic and localized solutions, the mass integral $M = \int_{-\infty}^{\infty} u(x, t) dx$ is zero [20].
2. There are no solitary waves on a constant background ("antisoliton theorem") [2, 17].
3. In the long-wave case ($\beta = 0$) there exists a family of stationary periodic waves including a limiting wave consisting of parabolic pieces [20].
4. An initial KdV soliton attenuates due to radiation of long small-amplitude waves and disappears as a whole entity in a finite time (the "terminal damping") [9].

Here we outline the recent results related to the behavior of a soliton and a pair of solitons interacting with a low-frequency "pump" wave, which compensates radiation losses and thus allows the existence of solitary waves. As will be shown below, the solitary waves on a background of a long carrier wave reveal a non-trivial dynamics.

3 Solitons on a Long Wave

3.1 Asymptotic Equations

For the sake of convenience, we use Eq. (2.1) in the dimensionless form:

$$(u_t + uu_x + u_{xxx})_x = u. \tag{3.1}$$

Note first that for a long wave one can neglect the term with the third derivative. The resulting "reduced" equation is integrable [11]. Its stationary solutions u_1 depending on one variable $S = x - Vt$ satisfy the equation

$$\frac{d^2}{dS^2}\left(\frac{1}{2}u_1^2 - c_1 u_1\right) = u_1. \tag{3.2}$$

Periodic solutions of this equation are shown in Fig. 2 (a detailed analysis of both periodic and solitary solutions of (3.2) can be found in Refs. [23, 25]). This family of periodic solutions includes, in particular, a limiting solution with a parabolic profile:

$$u_1(S) = \frac{1}{6}\left(S^2 - \frac{L^2}{12}\right), \quad -\frac{L}{2} \le S \le \frac{L}{2}, \quad c = \frac{L^2}{36}. \tag{3.3}$$

It represents a sequence of parabolic arcs one period of which is shown by line 3 in Fig. 2.

Assume now that one or more solitons close to the solitary solution of the KdV equation exist on the background of the long wave described by Eq. (3.3). In the present notation, a KdV soliton, which is a particular solution of Eq. (3.1) with the zero right-hand side, has the well-known form:

$$u_2 = A \operatorname{sech}^2 \frac{\zeta - S}{\Delta} - p, \tag{3.4}$$

Fig. 2 Samples of periodic solutions of Eq. (3.2) with different amplitudes

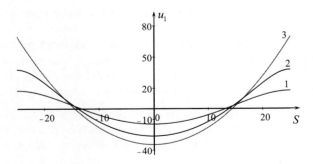

$$\text{where } \zeta = x - \int_0^t V \, dt, \quad V = \frac{A}{3} - p + u_1(\zeta), \quad \Delta = \sqrt{\frac{12}{A}}, \quad p = \frac{4\sqrt{3A}}{\Lambda}.$$

$$(3.5)$$

The small parameter p (the "pedestal") is added here to satisfy the zero mass condition mentioned above. In a more general case when N solitons are considered the solution u_2 reads:

$$u_2 = \sum_1^N u_{2n} = \sum_1^N \left(A_n \operatorname{sech}^2 \frac{\zeta - S_n}{\Delta_n} - p_n \right), \qquad (3.6)$$

where the parameters of each soliton are linked by the relations (3.5) with the corresponding local values of u_1.

Now let us seek for a solution to Eq. (3.1) in the form $u = u_1 + u_2$ where u_1 is given by Eq. (3.3) and make natural assumptions that $A_n \gg \max(u_1)$ and $\Delta \ll L$, i.e., each soliton is much narrower than the background wavelength L, but its peak value is much greater than that of the background wave. Substituting the corresponding expressions into Eq. (3.1) and separating the "fast" motions associated with the solitons from the "slow" motion in the long background wave, one obtains an equation for the soliton field:

$$\frac{\partial u_{2n}}{\partial t} + u_{2n} \frac{\partial u_{2n}}{\partial x} + \frac{\partial^3 u_{2n}}{\partial x^3} = -\frac{\partial}{\partial x} \left(u_1 u_{2n} + \sum_{m \neq n} u_{2m} u_{2n} \right). \qquad (3.7)$$

Here u_{2m} are represented by their "tails" as mentioned above.

From here, equations for the coordinates and amplitudes of solitons can be obtained, as it was derived by the authors in Ref. [22] for a single soliton on a long wave. It can be done either by rigorous application of the asymptotic method (see, e.g., [9, 21]) or simply by multiplying Eq. (3.7) by each term in the sum (3.6) and integrating over x in the vicinity of each soliton.

In general, there is an interplay between three processes: (i) interaction between the solitons considered as the effect of an exponential asymptotic ("tail") of one soliton on another; (ii) interaction of each of them with the long-wave background "pump"; and (iii) radiation losses of solitons. Here we describe the case of two solitons interacting with the parabolic wave (3.3), for which the following fourth-order system of ODEs can be obtained:

$$\begin{cases} \frac{dS_1}{dt} = \frac{A_1}{3} + \frac{1}{6}\left(S_1^2 - \frac{L^2}{12}\right) + A_2 \exp\left(-\sqrt{\frac{A_2}{3}}|S_2 - S_1|\right) - \frac{L^2}{36}, \\ \frac{dS_2}{dt} = \frac{A_2}{3} + \frac{1}{6}\left(S_2^2 - \frac{L^2}{12}\right) + A_1 \exp\left(-\sqrt{\frac{A_1}{3}}|S_2 - S_1|\right) - \frac{L^2}{36}, \\ \frac{dA_1}{dt} = -4\sqrt{3}A_1 - \frac{4}{9}A_1 S_1 - \frac{16}{3}A_1 A_2 \sqrt{\frac{A_2}{3}} \exp\left(-\sqrt{\frac{A_2}{3}}|S_2 - S_1|\right), \\ \frac{dA_2}{dt} = -4\sqrt{3}A_2 - \frac{4}{9}A_2 S_2 + \frac{16}{3}A_1 A_2 \sqrt{\frac{A_1}{3}} \exp\left(-\sqrt{\frac{A_2}{3}}|S_2 - S_1|\right). \end{cases} \qquad (3.8)$$

First we mention two particular cases considered earlier.

3.2   Interaction of Solitons Without Rotation

This is the case when Eq. (3.1) becomes the classic KdV equation and the set (3.8) is reduced to a simpler system:

$$\begin{cases} \frac{dS_1}{dt} = \frac{A_1}{3} + A_2 \exp\left(-\sqrt{\frac{A_2}{3}}|S|\right), \\ \frac{dS_2}{dt} = \frac{A_2}{3} + A_1 \exp\left(-\sqrt{\frac{A_1}{3}}|S|\right), \\ \frac{dA_1}{dt} = -\frac{16}{3}A_1 A_2 \sqrt{\frac{A_2}{3}} \exp\left(-\sqrt{\frac{A_2}{3}}|S|\right), \\ \frac{dA_2}{dt} = \frac{16}{3}A_1 A_2 \sqrt{\frac{A_1}{3}} \exp\left(-\sqrt{\frac{A_2}{3}}|S|\right). \end{cases} \qquad (3.9)$$

Here $S = S_1 - S_2$ is the distance between solitons. The corresponding asymptotic theory of this process was earlier developed for solitons with close amplitudes [7] (see also [21]) Comparison with the exact solution confirms the validity of Eq. (3.9). The main result is illustrated above in Fig. 1.

3.3   A Single Soliton on a Long Wave with Rotation

As mentioned above, this problem was recently considered in [22]. The corresponding equations for soliton position S and amplitude A are a particular case of the above set of Eq. (3.8) when $A_2 = 0$:

$$\begin{cases} \frac{dS}{dt} = \frac{A}{3} + \frac{1}{6}\left(S^2 - \frac{L^2}{12}\right) - \frac{L^2}{36}, \\ \frac{dA}{dt} = -4\sqrt{3}A - \frac{4}{9}AS. \end{cases} \qquad (3.10)$$

Figures 3 and 4 illustrate the behavior of a soliton on the parabolic wave. Figure 3

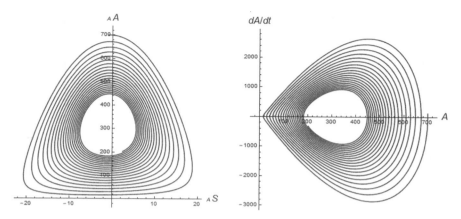

Fig. 3 Phase plane of Eq. (3.10) for $0 \leq t \leq 20$ in the variables (A, S) (left panel) and $(dA/dt, A)$ (right panel). The initial condition is: $S(0) = -1$, $A(0) = 700$. The wave period is $L = 50$

Fig. 4 Time dependence of soliton position $S(t)$ in the time interval $60 \leq t \leq 100$

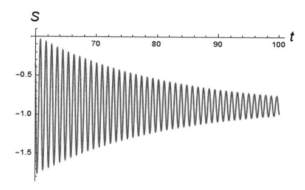

shows the phase plane of Eq. (3.10) in terms of the dependences $A(S)$ and $(dA/dt, A)$.

Note that the initial, large-amplitude oscillations are strongly nonlinear and close to the separatrix, as seen on the right panel. The equilibrium position of the soliton is always located at $S < 0$, where the long wave produces a positive work that compensates the radiation losses. This is illustrated in Fig. 4 which shows variation of soliton position $S(t)$ with respect to the minimum of the parabolic function u_2 at a larger time interval, $60 \leq t \leq 100$. In Ref. [22] the analytical solution for the equilibrium values of S and A was obtained.

3.4 Two Solitons on a Parabolic Background

Now we return to the general system (3.8) for two solitons on a parabola. As mentioned, here an interplay between such factors as interaction of solitons, their radiation losses, and interaction with the background takes place. These factors have their own time scales: the losses are a slow process taking many periods of oscillation in our case; on the contrary, interaction of solitons occurs fast during the oscillations on the background shown above for a single soliton. Here again, we distinguish between two cases: overtaking interaction for the initially strongly different solitons, and exchange interaction for solitons having close amplitudes from the very beginning.

3.4.1 Overtaking Interaction

Figure 5 illustrates the former case. Interpreting this behavior, one should remember that due to overlapping, the frontal and rear solitons are periodically transposed. When the solitons approach each other closely enough, they quickly exchange positions. (see almost vertical yellow lines in Fig. 5a).

Radiation losses take their effect at a longer time, as seen in Fig. 6. This process tends to an equilibrium in which both solitons acquire the same amplitudes, but

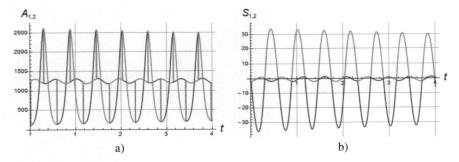

Fig. 5 Overlapping interaction of solitons. Initial conditions are $S_1 = -2$, $S_2 = 2$, $A_1 = 1250$, $A_2 = 125$. Dynamics of soliton amplitudes (**a**) and phases (**b**) at the initial stage of the process. Small oscillations pertain to the large soliton, large oscillations—to the small soliton

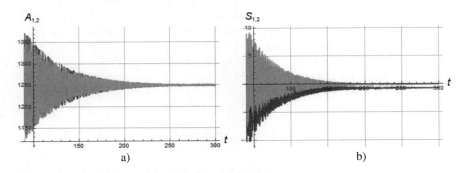

Fig. 6 Evolution of soliton amplitudes and positions at a long time interval, $0 \leq t \leq 300$

Fig. 7 The $(A_{1,2}, S)$ projection of the phase space of Eq. (3.8) for the initial time interval $0 \leq t \leq 10$ in the case of overtaking interaction. The left part of the plot correspond to A_1, the right part—to A_2. A small shift of symmetry to the left can be seen here

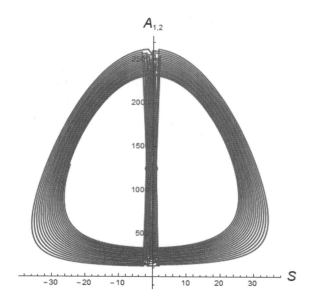

slightly different positions on the parabolic wave; the latter indicates that the interaction remains significant even asymptotically. An illustrative pattern of interplay between the solitons can be seen on the projections of 4D phase space of Eq. (3.8) as shown in Fig. 7. This figure shows that the exchange interaction occurs near the pump minimum, $S = 0$, but with a small shift to the left, where the pump works over solitons. Moreover, eventually soliton amplitudes become close to each other, so that the overlapping interaction transforms to that of the exchange type.

Figure 8 shows the dynamics of solitons (3.4) with the amplitudes and positions calculated from Eq. (3.8).

Finally, Figure 9 shows evolution of the Fourier spectrum of soliton amplitude A_1 at different time intervals of 30 units each.

Here, the fundamental harmonic is $f = 13.6$ in the dimensionless frequency units (see vertical dashed lines), and the spectrum in its vicinity is narrow, except for the first time interval where the spectrum is wider and it is slightly downshifted. As one can see from this figure, the fundamental harmonic quickly stabilizes at $f = 13.6$, whereas the higher frequency band shifts to the right, decreases with respect to the main frequency peak, and eventually disappears. The fundamental frequency in these spectra corresponds to the decaying oscillations of each soliton near the trough of the parabolic wave. The higher frequency bands reflect a more complex dynamics, which includes interaction between the solitons occurring faster than their individual oscillations on the parabola, see Fig. 5. The shift of this band towards higher frequencies apparently happens because the distance between solitons decreases with time.

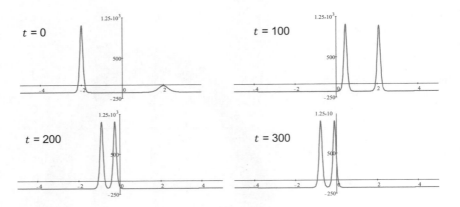

Fig. 8 Solitons with the amplitudes and positions on the parabola numerically calculated from Eq. (3.8). Only a fragment is shown of the total length $L = 100$ in the vicinity of the parabola's trough (so that the parabolic background looks like a straight line in this scale)

3.4.2 Exchange Interaction

Now consider the exchange interaction of solitons with amplitudes close to each other from the very beginning. Figure 10 shows the initial stage of this process. In this case, soliton amplitudes vary by less than 10% during a fast collision, and their trajectories do not intersect at any time. Asymptotically soliton amplitudes become equal, but, as above, their positions remain different, both at the decreasing slope of the pump, as shown in Fig. 11.

The phase space projection now looks like that shown in Fig. 12. For the most time the phase trajectories remain separated, so that the solitons interact at short intervals when they are close to each other. The shift of the corresponding area towards negative S is stronger in this case and during the same time interval ($0 < t < 10$) the trajectories fill the area much denser; the latter is due to a smaller oscillation amplitude and a faster exchange.

Figure 13 shows the amplitude spectra at different time intervals. Again, the basic oscillations are due to interaction with the long wave. In this case the higher frequency bands are expressed better than in Fig. 9, and the second high-frequency band was also generated (not shown in the plot). This is in agreement with Fig. 10 where a complex modulation of amplitudes and phases of solitons is clearly seen.

One of the interesting general features of the dynamics of solitons trapped near the trough of a long wave is that even if their initial amplitudes strongly differ, their amplitudes eventually become close to each other so that their long-time interaction always tends to an exchange type.

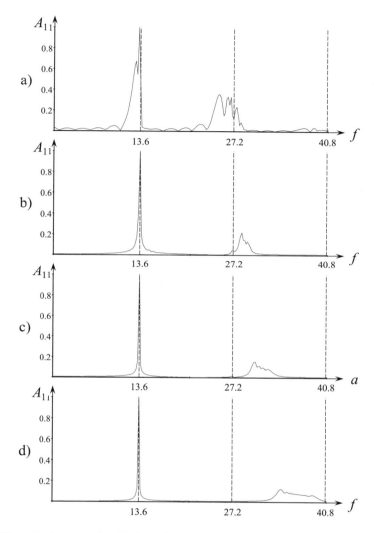

Fig. 9 Normalized spectra of oscillations of soliton amplitude A_1 at different time intervals for the overlapping interaction. From top to bottom: **a** $-0 \leq t \leq 30$; **b** $-30 \leq t \leq 60$; **c** $-60 \leq t \leq 90$; **d** $-90 \leq t \leq 120$

4 Conclusions

Here we have outlined the dynamics of solitons interacting with each other and with a long-wave background. The asymptotic procedure reduces the initial non-integrable PDE to the system of homogeneous ODEs with variable coefficients describing the amplitudes and trajectories of solitons as radiating particles. The resulting motion is

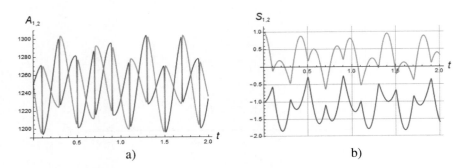

Fig. 10 Exchange interaction of solitons. Dynamics of soliton amplitudes (**a**) and phases (**b**) at the initial stage of the process. Initial conditions are $S_1 = -1$, $S_2 = 1$, $A_1 = 1250$, $A_2 = 1250$

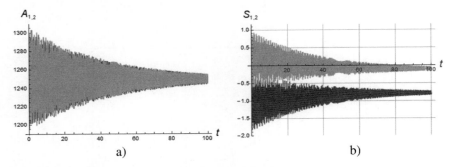

Fig. 11 Evolution of soliton amplitudes (**a**) and positions (**b**) at a longer time interval, $0 \leq t \leq 100$

Fig. 12 Same as in Fig. 7, but for the exchange interaction. Here the axis $S \approx -0.5$ divides the positions of solitons, which interact at around this position

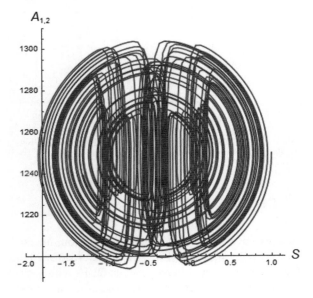

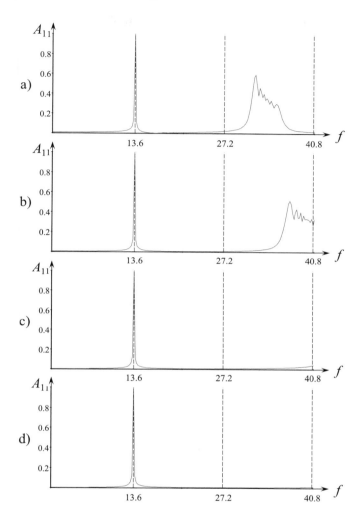

Fig. 13 Normalized spectra of oscillations of soliton amplitude A_1 in different time intervals for the exchange interaction. From top to bottom: **a** $-0 \le t \le 30$; **b** $-30 \le t \le 60$; **c** $-60 \le t \le 90$; **d** $-90 \le t \le 120$

due to the interplay of different factors: interaction between the solitons, which affect each other by their "tails", dissipation due to radiation of each soliton, and energy exchange with the long wave. It is often possible to distinguish these processes by their time scales. The final stage is always an equilibrium (here a non-zero equilibrium is of interest); it is noteworthy that for two solitons the amplitudes tend to the same value, whereas their phases remain different, so that the solitons remain interacting with each other even asymptotically.

There is a field for a further development of the model. First, the case of three and more solitons can supposedly result in chaotic motions, similarly to that shown in [5] for electromagnetic solitons. Second, to apply the above results to real physical situations one may need to consider additional factors making the dynamics even more complex. One of them is the action of solitons on the long wave, which was assumed given in the above consideration. For a single soliton this effect has been considered in [22]. Another factor is the action of the field radiated by the frontal soliton on the rear one. These factors can be included into the adiabatic theory, but due to the complexity of the situation, the result should be compared with a direct numerical simulation of the basic Eq. (2.1). Our intention here was to demonstrate that, along with the possible physical applications, the classical concept of solitons as particles can be extended far enough to form a new and interesting part of nonlinear dynamical science.

Appendix

Since this edition is dedicated to the memory of Valentin Afraimovich, it is worth saying that both authors have been in friendly interactions with him for a long time, and one of us (L. Ostrovsky) had a pleasure to be a co-editor of the journal "Discontinuity, Nonlinearity and Complexity" with Valentin. Note that this paper is conceptually close to the research interests of Valentin who was a world-renowned expert and the author of several books in the field of complex dynamics. Below there are two photos of him with the authors of this paper.

V. Afraimovich (left) and L. Ostrovsky in Xi'an, China, August 2014.

V. Afraimovich (right) and Y. Stepanyants at the conference in San Diego, USA in April, 2011.

References

1. Ablowitz MJ, Segur H (1981) Solitons and the inverse scattering transform. SIAM, Philadelphia
2. Galkin VM, Stepanyants Yu A (1991) On the existence of stationary solitary waves in a rotating fluid. Prikl Matamat i Mekhanika 55(6):1051−1055 (in Russian. Engl. transl.: J Appl Maths Mechs 1991 55(6):939−943)
3. Gerkema T, Zimmerman JTF (1995) Generation of nonlinear internal tides and solitary waves. J Phys Oceanogr 25:1081–1094
4. Gorshkov KA, Ostrovsky LA, Papko VV (1976) Interactions and bound states of solitons as classical particles. Zh Eksp Teor Fiz 71(2):585−593 (in Russian. Engl. transl.: Sov Phys JETP 1976 44(2):306−311)
5. Gorshkov KA, Ostrovsky LA, Papko VV (1977) Soliton turbulence in the system with weak dispersion. DAN SSSR 235(1):70–73 (in Russian)
6. Gorshkov KA, Ostrovsky LA, Papko VV, Pikovsky AS (1979) On the existence of stationary multisolitons. Phys Lett A 74(3–4):177–179
7. Gorshkov KA, Ostrovsky LA (1981) Interaction of solitons in nonintegrable systems: direct perturbation method and applications. Physica D 3:428–438
8. Gorshkov KA, Pelinovsky DE, Stepanyants Yu A (1993) Normal and anomalous scattering, formation and decay of bound-states of two-dimensional solitons described by the Kadomtsev-Petviashvili equation. Zh Eksp Teor Fiz 104(2):2704−2720 (in Russian. Engl. transl.: Sov Phys JETP 1993 77(2):237–245)
9. Grimshaw RHJ, He J-M, Ostrovsky LA (1998) Terminal damping of a solitary wave due To radiation in rotational systems. Stud Appl Math 101:197–210
10. Grimshaw R, Helfrich K (2012) The effect of rotation on internal solitary waves. IMA J Appl Math 77:326–339
11. Grimshaw RHJ, Helfrich K, Johnson ER (2012) The reduced Ostrovsky equation: integrability and breaking. Stud Appl Math 129:414–436
12. Grimshaw RHJ, Helfrich KR, Johnson ER (2013) Experimental study of the effect of rotation on nonlinear internal waves. Phys Fluids 25:056602
13. Grimshaw R, da Silva J, Magalhaes J (2017) Modelling and observations of oceanic nonlinear internal wave packets affected by the Earth's rotation. Ocean Model 16:146–158
14. Karpman VI, Maslov EM (1977) Perturbation theory for solitons. Zh Eksp Teor Fiz 73:537–559 (in Russian. Engl. transl.: Sov Phys JETP 1977, 46:281–295)
15. Kaup DJ, Newell AC (1978) Soliton as particles, oscillator and in slowly changing media: a singular perturbation theory. Proc Roy Soc London A 301(1701):413–446
16. Keener JP, McLaughlin DW (1977) Soliton under perturbation. Phys Rev A 16:777–790
17. Leonov AI (1981) The effect of Earth rotation on the propagation of weak nonlinear surface and internal long oceanic waves. Ann New York Acad Sci 373:150–159
18. Li Q, Farmer DM (2011) The generation and evolution of nonlinear internal waves in the deep basin of the South China Sea. J Phys Oceanogr 41:1345–1363
19. Matveev VB, Salle MA (1991) Darboux Transformations and Solitons. Springer
20. Ostrovsky LA (1978) Nonlinear internal waves in a rotating ocean. Okeanologia 18:181−191 (in Russian. Engl. transl.: Oceanology 18(2):119−125)
21. Ostrovsky LA (2015) Asymptotic perturbation theory of waves. Imperial College Press
22. Ostrovsky LA, Stepanyants YA (2016) Interaction of solitons with long waves in a rotating fluid. Physica D 333:266–275
23. Parkes EJ (2007) Explicit solutions of the reduced Ostrovsky equation. Chaos Solitons Fractals 31:602–610
24. Ramirez C, Renouard D, Stepanyants YuA (2002) Fluid Dyn Res 30(3):169–196
25. Stepanyants YA (2006) On stationary solutions of the reduced Ostrovsky equation: periodic waves, compactons and compound solitons. Chaos Solitons Fractals 28:193–204
26. Zabusky NJ, Kruskal MD (1965) Interaction of "solitons" in a collisionless plasma and the recurrence of initial states. Phys Rev Lett 15(6):240–243

A Climate-Economy Model with Endogenous Carbon Intensity

Dmitry V. Kovalevsky

Abstract We develop a simple aggregate nonlinear model of global coupled climate-socioeconomic system and explore with model simulations possible futures with and without climate mitigation actions. Carbon emissions produced by the global economy cause global mean temperature to rise that, in its turn, leads in the proposed modelling framework to increasing depreciation rate of physical capital slowing down or even reverting the economic growth. In a particular case of Business-as-Usual scenario, it is possible to derive exact analytical solution of the reduced model. Dependent on model parameters, Business-as-Usual scenario leads either to a collapse of modelled economy or, at best, to stagnation, while the mitigation scenario is characterized by sustainable growth of decarbonised economy at an affordable cost of mitigation actions.

1 Introduction

Any attempt to model processes in which humans are involved, from individuals to societies to coupled socio-natural systems, immediately brings researcher to the realm of strongly nonlinear phenomena [1, 2, 12, 31, 33, 34]. This is particularly true for integrated assessment modelling of coupled climate-socioeconomic dynamics under conditions of adverse climate change [4, 5, 7–11, 13, 21, 23, 25]. With many substantial negative impacts projected even for moderate climate scenarios, and much more dramatic impacts expected for high-end scenarios [14, 17, 29, 32, 36], the problem of global warming calls for mitigation actions to avoid long-term impacts of climate change, and also for adaptation measures in short- and mid-term informed by climate services [15, 20, 30].

While detailed assessment of mitigation measures requires complex and often highly disaggregated models, certain basic properties of coupled climate-socioeconomic

D. V. Kovalevsky (✉)
Climate Service Center Germany (GERICS), Helmholtz-Zentrum Hereon, Fischertwiete 1, 20095 Hamburg, Germany
e-mail: dmitrii.kovalevskii@hereon.de

dynamics can be easily understood already with simple aggregate conceptual models. Therefore, in this book chapter we develop, primarily for didactic purposes, a simple nonlinear climate-economy model to project possible futures of the world dynamics with and without climate mitigation actions.

The rest of this book chapter is organised as follows. In Sect. 2, we describe a model of coupled climate-socioeconomic system. With this model, we explore the Business-as-Usual (BAU) scenario (under assumption of no climate mitigation actions) in Sect. 3, and later the mitigation scenario in Sect. 4. Section 5 concludes.

2 Model Description

In another chapter of this book, we presented a model family for regional economy affected by climate change [24], where regional economic dynamics were driven by exogenous climate forcing, and there was no feedback from economic system to climate system. This approach is justified for modelling regional and local climate impacts. However, when modelling the dynamics of coupled climate-economy system at a global scale, the architecture of the model should be fundamentally different. Therefore, we now adopt a closed-loop model design where the global economy affects the global climate system through carbon emissions, while changing global climate causes climate damages to the global economic system (Fig. 1).

The dynamics of the global economy will be described on the basis of the AK model of economic growth [3, 19, 26, 27]. In the AK model, the output Y (world GDP) is assumed to be proportional to capital K,

$$Y = AK, \tag{1}$$

where $A = $ const. is the technology parameter. A constant fraction s of the output (called the savings rate, $0 < s < 1$) is invested in capital, the remainder $(1 - s)$ is consumed. The capital depreciates at a rate δ. Overall, this leads to capital dynamics

Fig. 1 A flow chart of the coupled climate-economy model

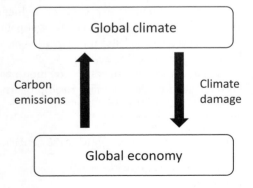

described by a simple ordinary differential equation (ODE)

$$\dot{K} = (sA - \delta)K. \tag{2}$$

If the savings rate and the depreciation rate are kept constant ($s = $ const., $\delta = \delta_0 = $ const.), and

$$r_0 = sA - \delta_0 \tag{3}$$

is positive, then the economy exhibits exponential growth, r_0 being the constant growth rate.

However, the adverse climate change would slow down the economic growth. Generally, impacts of climate change might be introduced in the model (2) and in similar models in two ways. Either a climate damage function $d(T)$ dependent on global mean temperature increase above the pre-industrial level T (referred as temperature for brevity below) is included as a measure of reduction of global output[1] [7, 25, 28, 35],

$$Y_{\text{eff}} = (1 - d(T))\,Y, \tag{4}$$

or the depreciation rate δ is assumed to be growing with temperature: $\delta = \delta(T)$ [6, 16, 22]. While the former approach is more broadly used, we will stick to the latter one and assume that the depreciation rate grows linearly with temperature:

$$\delta(T) = \delta_0\,(1 + \varepsilon(T - T_0))\,, \tag{5}$$

where ε is a constant sensitivity of depreciation rate to temperature change and T_0 is the initial temperature value at the start year of simulations.

By substituting (5) in (2) and making use of (3), we get a modified equation for capital dynamics

$$\dot{K} = [r_0 - \delta_0\varepsilon(T - T_0)]\,K. \tag{6}$$

Note that (6) can be also interpreted in terms of climate damages to output, like in (4),

$$\dot{K} = [s\,(1 - d(T))\,A - \delta_0]\,K, \tag{7}$$

if we introduce the climate damage function of the form

$$d(T) = \frac{\delta_0\varepsilon}{sA}\,(T - T_0) \tag{8}$$

while keeping the depreciation rate in (7) constant ($\delta = \delta_0$). We note, however, that while climate damage functions linear in temperature are sometimes used in inte-

---

[1] In more advanced models, climate damage functions can also depend on additional climate parameters as sea level rise, cf. [28].

grated assessment models [18], nonlinear (and often—strongly nonlinear) parameterizations are much more broadly adopted [28, 37].

The temperature dynamics are driven by carbon emissions E that are assumed to be proportional to the global output,

$$E = \eta Y, \tag{9}$$

where η is the carbon intensity of the global economy. In Sect. 3 we will consider in detail the Business-as-Usual (BAU) scenario with constant η ($\eta = \eta_0 = $ const), while in Sect. 4 we will explore the climate mitigation scenario with η endogenously decreasing as a result of purpose-oriented investment in renewable energy.

To describe the temperature dynamics, we use a simple linear ODE

$$\dot{T} = \xi E - \lambda_T T \tag{10}$$

with the emissions forcing and the relaxation term, where ξ is the temperature sensitivity to emissions and λ_T is the relaxation rate. In Appendix, we compare (10) with and calibrate it against a more complex nonlinear model (31a)–(31b), to determine the numerical values of model parameters ξ and λ_T.

In case of BAU scenario with $\eta = \eta_0 = $ const, two ODEs (6) and (10) supplemented with auxiliary identities (1) and (9) and initial conditions (K_0, T_0) for the start year of simulations form the closed model to be analysed in the next Section. To explore the mitigation scenario, the model should be further supplemented with an ODE for carbon intensity dynamics. This will be done in Sect. 4.

3   Business-as-Usual Scenario

In the present Section, we explore in detail the BAU scenario with no mitigation actions and, therefore, with a constant carbon intensity ($\eta_0 = $ const.) As discussed above, in the BAU case the climate-economy model takes the form

$$\dot{K} = [r_0 - \delta_0 \varepsilon (T - T_0)] K, \tag{11a}$$

$$\dot{T} = \xi \eta_0 A K - \lambda_T T. \tag{11b}$$

In Sect. 3.1, we explore for didactic purposes the limiting case $\lambda_T = 0$, where an exact analytical solution of the system (11a)–(11b) can be obtained. We continue the discussion of properties of the derived analytical solution in Sect. 3.2, where we also assign the appropriate numerical values to model parameters. A general case of $\lambda_T > 0$ is analysed numerically in Sect. 3.3.

3.1 Analytical Solution in a Particular Case

The limiting case $\lambda_T = 0$ corresponds to modelling the climate system with extremely large values of its climate time scale (a question of model time scales is briefly tackled in the Appendix). We are primarily interested to explore in detail the case $\lambda_T = 0$ because it will be possible to derive analytical solution in a closed form for this particular case. At the same time, it should be mentioned that this case yields a substantially more dramatic socio-economic scenario than if λ_T were assigned a positive value reasonable from the point of view of climate science. Therefore in Sect. 3.3 we perform numerical simulations for the case $\lambda_T > 0$.

In case $\lambda_T = 0$, the dynamical system (11a)–(11b) takes the reduced form

$$\dot{K} = [r_0 - \delta_0 \varepsilon (T - T_0)] K, \tag{12a}$$
$$\dot{T} = \xi \eta_0 A K. \tag{12b}$$

To obtain the analytical solution of the system (12a)–(12b), we divide (12a) by (12b),

$$\frac{dK}{dT} = \frac{1}{\xi \eta_0 A} [r_0 - \delta_0 \varepsilon (T - T_0)], \tag{13}$$

and then integrate (13) over T. This yields

$$K(T) = K_0 + \frac{1}{\xi \eta_0 A} \left[r_0 (T - T_0) - \frac{\delta_0 \varepsilon}{2} (T - T_0)^2 \right]. \tag{14}$$

After substituting (14) in the temperature dynamics equation (12b), the latter takes the form

$$\dot{T} = \xi \eta_0 A K_0 + r_0 (T - T_0) - \frac{\delta_0 \varepsilon}{2} (T - T_0)^2. \tag{15}$$

Variables in (15) can be separated,

$$dt = -\frac{dT}{(\delta_0 \varepsilon / 2)(T - T_0)^2 - r_0 (T - T_0) - \xi \eta_0 A K_0}, \tag{16}$$

and after integrating both sides of (16) we get

$$t - t_0 = -\int_0^{T - T_0} \frac{d\Theta}{(\delta_0 \varepsilon / 2)\Theta^2 - r_0 \Theta - \xi \eta_0 A K_0}. \tag{17}$$

To perform integration in (17) explicitly, we introduce the notations

$$a = \frac{\delta_0 \varepsilon}{2}, \quad b = -r_0, \quad c = -\xi \eta_0 A K_0 \tag{18}$$

and

$$\Delta = b^2 - 4ac = r_0^2 + 2\delta_0\varepsilon\xi\eta_0 A K_0 \tag{19}$$

(note that $\Delta > 0$). Then the integral in the r.h.s. of (17) is of the form

$$\int \frac{dx}{ax^2 + bx + c} = \frac{1}{\sqrt{\Delta}} \ln\left|\frac{2ax + b - \sqrt{\Delta}}{2ax + b + \sqrt{\Delta}}\right| + C, \tag{20}$$

and (17) yields

$$t - t_0 = -\frac{1}{\sqrt{\Delta}} \ln\left|\frac{\delta_0\varepsilon(T - T_0) - r_0 - \sqrt{\Delta}}{\delta_0\varepsilon(T - T_0) - r_0 + \sqrt{\Delta}} \cdot \frac{\sqrt{\Delta} - r_0}{\sqrt{\Delta} + r_0}\right|. \tag{21}$$

To obtain from (21) an explicit expression for temperature, we define an auxiliary function of time

$$\varphi(t - t_0) = \frac{1 - \exp\left(-\sqrt{\Delta}(t - t_0)\right)}{(\sqrt{\Delta} - r_0) + (\sqrt{\Delta} + r_0)\exp\left(-\sqrt{\Delta}(t - t_0)\right)}. \tag{22}$$

Note that $\varphi(0) = 0$ and

$$\varphi(+\infty) = \frac{1}{\sqrt{\Delta} - r_0}. \tag{23}$$

Then, explicitly,

$$T(t) - T_0 = \frac{\Delta - r_0^2}{\delta_0\varepsilon}\varphi(t - t_0). \tag{24}$$

Note that, in view of (23), the asymptotic temperature increase above the initial value is positive:

$$T(+\infty) - T_0 = \frac{\sqrt{\Delta} + r_0}{\delta_0\varepsilon}. \tag{25}$$

In view of (19), the solution for temperature (24) can be also re-written in the form

$$T(t) = T_0 + 2\xi\eta_0 A K_0 \cdot \varphi(t - t_0). \tag{26}$$

Now we can substitute (26) in (14), that yields the solution for capital

$$K(t) = K_0\left[1 + 2r_0\varphi(t - t_0) - 2\delta_0\varepsilon\xi\eta_0 A K_0 \cdot \varphi^2(t - t_0)\right]. \tag{27}$$

Note that, in view of (25), $K(+\infty) = 0$.

Expressions (26)–(27) for temperature and capital, with the auxiliary function $\varphi(t - t_0)$ defined in (22), provide the exact analytical solution for the reduced model (12a)–(12b). Its properties will be discussed in more detail in the next Section.

3.2 Model Parameters and Properties of Analytical Solution

To plot the analytical solution (26)–(27) of the reduced model (12a)–(12b) (and, later, to perform simulations for the full model (11a)–(11b)), we have to specify the numerical values of model parameters and the initial conditions.

Consider for a while the full model (11a)–(11b). For our illustrative 'what-if' simulations, we choose the unperturbed growth rate of the global economy at a level of 4% per annum ($r_0 = 0.04\,\text{year}^{-1}$), although such a high growth rate has been seldom recorded in recent decades.[2] The unperturbed capital depreciation rate is chosen as $\delta_0 = 0.05\,\text{year}^{-1}$, a conventional choice of many modelling studies in economic growth [3]. Note that if we adopt the value $A = 0.4\,\text{year}^{-1}$ for the technology parameter [35], Eq. (3) would then yield the savings rate $s = 0.225$. The sensitivity ε in (11a) is chosen at a level $\varepsilon = 0.2\,(^\circ\text{C})^{-1}$; this means that the square bracket in (11a) would become equal to zero, and therefore the economy would have zero growth rate, when the temperature reaches the level

$$\bar{T} = T_0 + \frac{r_0}{\Delta\varepsilon} \tag{28}$$

being 4 °C higher than its initial value T_0.

Following the calibration procedure described in detail in Appendix, we choose the following values of parameters in (11b): $\xi = 0.0015$ K year/GtCO$_2$, $\lambda_T = 0.02\,\text{year}^{-1}$.

The start year of our simulations will be 2018 (model year 0). In BAU scenario, we assume that the observed value of η_0 for 2018 is kept constant over the entire horizon of simulations. We estimate $\eta_0 = E_0/Y_0$ as the ratio of global carbon emissions E_0 to world GDP Y_0 in 2018. The global CO$_2$ emissions from fossil fuels in 2018 were 36573 MtCO$_2$,[3] while the nominal world GDP in 2018 was \$85,804,390.60 million.[4] This yields $\eta_0 = 0.43$ GtCO$_2$/trlnUSD.

Regarding the initial conditions, we estimate the initial capital value $K_0 = Y_0/A$ by substituting the values for Y_0 in 2018 and for the parameter A provided above, that gives $K_0 = 214.5$ trlnUSD. Coming to the initial temperature value, as indicated in NOAA Global Climate Report for Annual 2018, the global mean temperature in 2018 was 0.97 °C above the 1880–1900 average, commonly used to represent the

---

pre-industrial conditions. In the three preceding years (2015–2017) the global mean temperature was more than 1 °C above the pre-industrial level.[5] Therefore, we choose the initial condition $T_0 = 1$ °C.

The analytical solution (26)–(27) of the reduced model is visualized in Fig. 2 (dashed curves). The dynamics of world GDP (or, equivalently, of the output Y proportional to capital (27), in accordance with the production function (1)) are shown in Fig. 2a, while the dynamics of temperature T (26) are plotted in Fig. 2b.

The BAU scenario under assumption $\lambda_T = 0$ is desperately pessimistic. As the climate time scale λ_T^{-1} is assumed to be infinite and therefore the temperature relaxation mechanism is deactivated, the temperature can only grow, converging in the long term to the limit (25). At large times, when $T \to T(+\infty)$, the growth rate of global economy (the square bracket in (12a)) converges to a negative limit, $r_0 - \delta\varepsilon(T - T_0) \to -\sqrt{\Delta} < 0$. Therefore, after a short period of growth, the global economy starts decaying and collapses due to adverse impacts of climate change.

3.3 Numerical Simulations in a General Case

We now return from the reduced model (12a)–(12b) to the full model (11a)–(11b) with $\lambda_T > 0$. For the full model, to derive an analytical solution is no longer possible, and the model dynamics can be explored with numerical simulations only. However, the model steady state (K^*, T^*) to which the dynamical system converges in the long term can still be easily found analytically. Indeed, (K^*, T^*) for which the r.h.s. of (11a)–(11b) becomes zero are equal to

$$K^* = \frac{\lambda_T}{\xi \eta_0 A}\left(T_0 + \frac{r_0}{\delta\varepsilon}\right), \qquad T^* = T_0 + \frac{r_0}{\delta\varepsilon}. \tag{29}$$

For numerical simulations, we choose the same values of model parameters as in Sect. 3.2 (with λ_T now equal to $0.02\,\text{year}^{-1}$). The full model dynamics are shown in Fig. 2 with solid curves. We see that the activation of temperature relaxation mechanism makes model dynamics somewhat less pessimistic than for the case $\lambda_T = 0$ considered in Sects. 3.1–3.2: the world GDP converges to a positive value, that is even higher than its initial value at the start of simulations, and the temperature growth is less dramatic than in the previous case (although still unacceptably high). However, keeping in mind the population growth projections[6] (that our simple model does not include at all), the GDP per capita would be rapidly decreasing, so the BAU scenario for the full model still gives too little reason for optimism.

---

[5] NOAA National Centers for Environmental Information, State of the Climate: Global Climate Report for Annual 2018, published online January 2019, retrieved on December 13, 2019.

[6] World Population Prospects 2019. The United Nations.

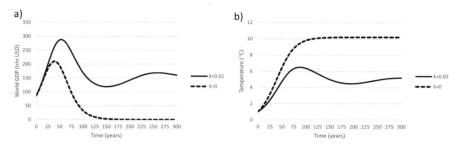

Fig. 2 Model dynamics of GDP (Panel a) and temperature (Panel b) under Business-as-Usual (BAU) scenario. Solid curve: $\lambda_T = 0.02\,\text{year}^{-1}$; dashed curve: $\lambda_T = 0$

4 Climate Mitigation Scenario

We now explore the climate mitigation scenario, with carbon intensity $\eta(t)$ in (9) endogenously decreasing due to investment in renewable energy. A certain fraction of world GDP (an affordable fraction, as will be seen from simulation results presented below) is now channelled in climate mitigation. We assume that this mitigation investment is proportional to the current value of carbon emissions $E(t) = \eta(t)Y(t)$ with a constant factor τ. Therefore, emissions are now priced globally – e.g. through a globally harmonized carbon tax of the rate τ. After the mitigation investment is made in year t, the remaining part of global output equal to $(1 - \tau\eta(t))\,Y(t)$ can be either 'conventionally' invested in physical capital or consumed. The partitioning between 'conventional' investment and consumption is $s: (1 - s)$, as before.

The rate of endogenous decrease of carbon intensity is assumed to be proportional to the mitigation investment with a constant factor β. The temperature dynamics equation takes the same form (10) as before. Overall, this leads to the now three-dimensional dynamical system

$$\dot{K} = [r_0 - \delta_0\varepsilon(T - T_0) - s\tau\eta A]\,K, \tag{30a}$$

$$\dot{T} = \xi\eta AK - \lambda_T T, \tag{30b}$$

$$\dot{\eta} = -\beta\tau\eta AK. \tag{30c}$$

For numerical simulations, we choose the same values of model parameters as for the BAU scenario. Additionally, we specify the carbon price τ at a constant level of 0.05 trlnUSD/GtCO$_2$ (equivalent to 50 USD/tCO$_2$), and choose $\beta = 0.0033$ GtCO$_2$/trlnUSD^2 as the decarbonisation parameter in (30c).

The simulated model dynamics for the mitigation scenario are shown in Fig. 3. As before, solid curves correspond to $\lambda_T = 0.02\,\text{year}^{-1}$, while dashed curves – to $\lambda_T = 0$. Panels a to c show the dynamics of three state variables of the model (the world GDP that is proportional to $K(t)$, temperature $T(t)$ and carbon intensity $\eta(t)$, respectively), while Panel d additionally visualizes carbon emissions $E(t)$.

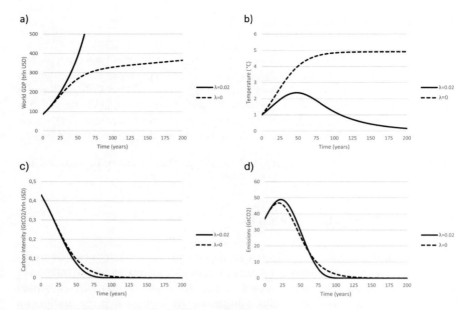

Fig. 3 Model dynamics of GDP (Panel a), temperature (Panel b), carbon intensity (Panel c) and carbon emissions (Panel d) under climate mitigation scenario. Solid curve: $\lambda_T = 0.02\,\text{year}^{-1}$; dashed curve: $\lambda_T = 0$

The world GDP (Fig. 3a) experiences rapid growth in case $\lambda_T = 0.02\,\text{year}^{-1}$ and moderate growth in case $\lambda_T = 0$. As before, in the limiting case $\lambda_T = 0$ the temperature converges to a positive value in the long term. However, in case $\lambda_T = 0.02\,\text{year}^{-1}$ the temperature dynamics are fundamentally different: after peaking, the temperature gradually returns to its pre-industrial level (Fig. 3b). This temperature relaxation to pre-industrial level is the reason for rapid economic growth in case $\lambda_T = 0.02\,\text{year}^{-1}$. A rapid decarbonisation of global economy is evident from Fig. 3c–d, where the dynamics of carbon intensity and emissions are shown, respectively (the related curves are similar for $\lambda_T = 0.02\,\text{year}^{-1}$ and $\lambda_T = 0$): both carbon intensity and emissions vanish in the long term, and the global economy becomes carbon-neutral. Regarding the cost of mitigation actions, the fraction of world GDP channelled in mitigation (equal to $\tau\eta(t)$, not shown in the figure as this quantity is proportional to $\eta(t)$ plotted in Fig. 3c) is equal to 2.15% at the start year of simulations and then rapidly and monotonically decreases, vanishing in the long term with the decarbonisation of global economy.

To summarise, the mitigation scenario yields substantially more optimistic dynamics than BAU, with sustainable growth of decarbonised economy in the long term at an affordable cost of mitigation actions.

5 Conclusions and Outlook

With a simple climate-economy model developed in the present book chapter, we explored both the Business-as-Usual (BAU) scenario with no climate mitigation action undertaken, and the mitigation scenario with carbon emissions priced globally and the revenues invested in endogenous reduction of carbon intensity. Expectedly, these two scenarios provide very different perspectives of world dynamics in the long term, with sustainable growth under the mitigation scenario as opposed to stagnation or even decay of global economy under BAU.

Both the global economy and the global climate system have been described in the model by very simple, not to say minimal, relations. However, even within this simple framework, the model dynamics are shown to be quite sensitive to parameters describing the global climate system, particularly to the value of temperature relaxation rate (or, equivalently, climate time scale).

The analysis of coupled climate-socioeconomic dynamics presented above was primarily focussed on the long term in relation to climate mitigation problem. In our future research, we are planning to address more closely the short- and mid-term dynamics with technically simple extensions of the developed model, to explore time horizons particularly relevant for planning and implementation of climate adaptation measures and the role of climate services in supporting adaptation.

Within the developed framework, adaptation actions might be introduced in the model e.g. as a new investment channel targeted to reduce the sensitivity ε of depreciation to temperature change. The decision on how much to invest in adaptation can be informed by climate services. The latter can be described in this simple modelling scheme as R&D actions aimed at more precise determination of parameters of temperature dynamics equation, also requiring a certain amount of investment.

Acknowledgements The reported study was supported by the Russian Foundation for Basic Research, research project No. 12-06-00381-a.

Appendix: Calibration of Linear Model for Temperature Dynamics

Below we calibrate the linear model for temperature dynamics represented by (10) against a nonlinear climate module used in [18] (see Eqs. (2.1)–(2.2) of the cited paper) to project the dynamics of coupled climate-economic system over millennial time scale:

$$\dot{\Theta} = \frac{1}{\tau_T} \left[\frac{\Delta T^*}{\ln 2} \ln \frac{C}{C_{PI}} - \Theta \right], \tag{31a}$$

$$\dot{C} = E - \frac{1}{\tau_C} (C - C_{PI}). \tag{31b}$$

In the nonlinear temperature dynamics equation (31a) we changed the notation for global mean temperature to Θ in order not to confuse its solution with the linear approximation provided by (10). The dynamics of atmospheric CO_2 concentration C driven by carbon emissions E are described by (31b). In (31a)–(31b) τ_C and τ_T is the characteristic carbon and climate time scale, respectively; C_{PI} is the pre-industrial atmospheric CO_2 concentration; ΔT^* is the equilibrium climate sensitivity (equal to the steady state temperature increase in response to a hypothetical scenario of CO_2 doubling from the pre-industrial level).

To calibrate (10) against (31a)–(31b), we first make the relaxation time scale in (10) equal to that in (31a); that means, we choose

$$\lambda_T = \frac{1}{\tau_T}. \tag{32}$$

We then choose ξ in (10) in such a way that the asymptotic temperature values in response to a constant emission forcing scenario would be close to each other for both models.

Suppose that $E(t) = E^* = \text{const}$. Then the asymptotic temperature value T^* corresponding to the steady state of (10) is

$$T^* = \frac{\xi}{\lambda_T} E^*. \tag{33}$$

For the model (31a)–(31b), the asymptotic CO_2 concentration value C^* is

$$C^* = C_{PI} + \tau_C E^*, \tag{34}$$

while the asymptotic temperature value Θ^* corresponding to C^* is

$$\Theta^* = \frac{\Delta T^*}{\ln 2} \ln \frac{C^*}{C_{PI}}, \tag{35}$$

or, explicitly,

$$\Theta^* = \frac{\Delta T^*}{\ln 2} \ln \left(1 + \frac{\tau_C}{C_{PI}} E^* \right). \tag{36}$$

If the term $\tau_C E^*/C_{PI}$ in (36) were small enough to justify an expansion of the form

$$\ln(1 + x) \sim x, \tag{37}$$

we would derive from (36)

$$\Theta^* \sim \frac{\Delta T^*}{\ln 2} \frac{\tau_C}{C_{PI}} E^*, \tag{38}$$

and, comparing with (33) and making use of (32), would choose

$$\xi' = \frac{\Delta T^*}{C_{PI} \ln 2}.$$ (39)

However, numerical simulations provided below show that condition

$$\frac{\tau_C}{C_{PI}} E^* \ll 1$$ (40)

necessary to justify the expansion (37) does not hold with enough accuracy for simulation scenarios reasonably close to reality. Therefore, we introduce in (39) the correction factor $q < 1$ (its value has to be defined from numerical simulations) and so replace (39) with an approximation

$$\xi = q \frac{\Delta T^*}{C_{PI} \ln 2}.$$ (41)

To evaluate q in (41), we make numerical simulations with (10) and with the model (31a)–(31b) from 2018 (model year 0) over the horizon of 300 years to the future under a hypothetical (and very implausible) scenario of annual carbon emissions kept constant at their 2018 level.

Following [18], we choose the values of model parameters $\tau_C = \tau_T = 50$ years (so $\lambda_T = 0.02 \, \text{year}^{-1}$ in (10)), $\Delta T^* = 3 \, \text{K}$, $C_{PI} = 280$ ppmv.

For our constant emission scenario we round the value of global CO_2 emissions from fossil fuels in 2018 provided in Sect. 3.2 down to $E^* = 36 \, \text{GtCO}_2/\text{year}$.

As discussed in Sect. 3.2, it is reasonable to choose the value $T_0 = 1 \, ^\circ\text{C}$ as the initial value for temperature increase above the pre-industrial level in 2018. Regarding the initial condition for CO_2 concentration, keeping in mind that the 2018 global annual mean marine surface CO_2 concentration was 407.38 ppmv,[7] we round this value down to $C_0 = 400$ ppmv.

As carbon emissions are conveniently expressed in $GtCO_2/\text{year}$, we will also need to convert ppmv to $GtCO_2$. 1 ppmv of atmosphere CO_2 is equivalent to 2.13 GtC,[8] and, as one GtC equals 3.67 $GtCO_2$,[9] 1 ppmv = 2.13 × 3.67 = 7.82 $GtCO_2$.

The simulated response of the linear model (10) and the nonlinear model (31a)–(31b) to the constant emissions scenario is shown in Fig. 4. The solid curve corresponds to the nonlinear solution $\Theta(t)$, while the dashed curve corresponds to the linear solution $T(t)$ with the parameter ξ' provided by (39)—numerically, its value is $\xi' = 0.00198 \, \text{K year/GtCO}_2$. We see that qualitatively both models yield similar dynamics; however, the asymptotic temperature values are quite different: $T^* = 3.56 \, ^\circ\text{C}$ in the linear model, which is almost one degree warmer than

---

[7] Globally averaged CO_2 marine surface annual mean data. Ed Dlugokencky and Pieter Tans, NOAA/ESRL.

[8] Carbon Dioxide Information Analysis Center—Conversion Tables.

[9] The Carbon Budget—Climate nexus.

Fig. 4 The simulated response of temperature dynamics models to the constant emissions scenario. Solid curve: nonlinear model (31a)–(31b); dashed curve: linear model (10) with $q = 1$; dotted curve: linear model (10) with $q = 0.75$ (see Appendix for a detailed discussion)

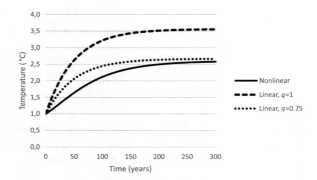

$\Theta^* = 2.60\,°C$ in the nonlinear model. This discrepancy is not surprising as the non-dimensional parameter in the r.h.s. of the criterion (40) is numerically equal to 0.822 under this scenario—not really much less than unity. Therefore, we use the adjusted value of ξ provided by (41) with the correction factor $q = 0.75$—numerically, $\xi = 0.00149\ \text{K year/GtCO}_2$. The corresponding linear solution is shown in Fig. 4 with the dotted curve, and we see that the asymptotic temperature values for both models now virtually coincide, while their transitional dynamics are pretty close. For further use in the coupled climate-economy model we round the adjusted value of ξ down to $\xi = 0.0015\ \text{K year/GtCO}_2$.

References

1. Afraimovich VS, Rabinovich MI, Varona P (2004) Heteroclinic contours in neural ensembles and the winnerless competition principle. Int J Bifurcation Chaos 14(04):1195–1208
2. Afraimovich VS, Zhigulin VP, Rabinovich MI (2004) On the origin of reproducible sequential activity in neural circuits. Chaos Interdiscip J Nonlinear Sci 14(4):1123–1129
3. Barro RJ, Sala-i Martin XI (2003) Economic Growth, 2nd edn. The MIT Press
4. Barth V (2003) Integrated assessment of climate change using structural dynamic models. PhD Thesis, Max-Planck-Institut für Meteorologie, Hamburg
5. Barth V, Hasselmann K (2005) Analysis of climate damage abatement costs using a dynamic economic model. Vierteljahresheft zur Wirtschaftsforschung (DIW) 74(2):148–163
6. Bretschger L, Valente S (2011) Climate change and uneven development. Scand J Econ 113(4):825–845
7. Calvin K, Bond-Lamberty B, Clarke L, Edmonds J, Eom J, Hartin C, Kim S, Kyle P, Link R, Moss R, McJeon H, Patel P, Smith S, Waldhoff S, Wise M (2017) The SSP4: a world of deepening inequality. Glob Environ Change 42:284–296
8. Capellán-Pérez I, González-Eguino M, Arto I, Ansuategi A, Dhavala R, Patel P, Markandya A (2014) New climate scenario framework implementation in the GCAM integrated assessment model. Working Papers 2014-04, BC3
9. Capellán-Pérez I, Arto I, Polanco-Martínez J, González-Eguino M, Neumann M (2016) Likelihood of climate change pathways under uncertainty on fossil fuel resource availability. Energy Environ Sci 9:2482–2496

10. Hallegatte S, Ghil M (2008) Natural disasters impacting a macroeconomic model with endogenous dynamics. Ecol Econ 68(1):582–592
11. Hasselmann K (2013) Detecting and responding to climate change. Tellus B: Chem Phys Meteorol 65(1):20,088
12. Hasselmann K, Kovalevsky DV (2013) Simulating animal spirits in actor-based environmental models. Environ Modell Softw 44:10–24
13. Hasselmann K, Cremades R, Filatova T, Hewitt R, Jaeger C, Kovalevsky D, Voinov A, Winder N (2015) Free-riders to forerunners. Nat Geosci 8:895–898
14. Hoegh-Guldberg O, Jacob D, Taylor M, Guillén Bolaños T, Bindi M, Brown S, Camilloni IA, Diedhiou A, Djalante R, Ebi K, Engelbrecht F, Guiot J, Hijioka Y, Mehrotra S, Hope CW, Payne AJ, Pörtner HO, Seneviratne SI, Thomas A, Warren R, Zhou G (2019) The human imperative of stabilizing global climate change at 1.5 °C. Science 365(6459)
15. van den Hurk BJ, Bouwer LM, Buontempo C, Döscher R, Ercin E, Hananel C, Hunink JE, Kjellström E, Klein B, Manez M, Pappenberger F, Pouget L, Ramos MH, Ward PJ, Weerts AH, Wijngaard JB (2016) Improving predictions and management of hydrological extremes through climate services
16. Ikefuji M, Horii R (2012) Natural disasters in a two-sector model of endogenous growth. J Publ Econ 96(9):784–796
17. Jacob D, Kotova L, Teichmann C, Sobolowski SP, Vautard R, Donnelly C, Koutroulis AG, Grillakis MG, Tsanis IK, Damm A, Sakalli A, van Vliet MTH (2018) Climate impacts in Europe under +1.5°C global warming. Earth's Future 6(2):264–285
18. Kellie-Smith O, Cox PM (2011) Emergent dynamics of the climate–economy system in the Anthropocene. Philos Trans R Soc A: Math Phys Eng Sci 369(1938):868–886
19. Knight FH (1944) Diminishing returns from investment. Journal of Political Economy 52:26–47
20. Koldunov NV, Kumar P, Rasmussen R, Ramanathan A, Nesje A, Engelhardt M, Tewari M, Haensler A, Jacob D (2016) Identifying climate change information needs for the Himalayan region: results from the GLACINDIA stakeholder workshop and training program. Bull Am Meteorol Soc 97(2):ES37–ES40
21. Kovalevsky DV (2014) Balanced growth in the Structural Dynamic Economic Model SDEM-2. Discontinuity, Nonlinearity, and Complexity 3(3):237–253
22. Kovalevsky DV (2016) Exact analytical solutions of selected behaviourist economic growth models with exogenous climate damages. Discontinuity, Nonlinearity Complex 5(3):251–261
23. Kovalevsky DV (2016) Introducing increasing returns to scale and endogenous technological progress in the Structural Dynamic Economic Model SDEM-2. Discontinuity, Nonlinearity Complex 5(1):1–8
24. Kovalevsky DV, Máñez Costa M (2020) Dynamics of water-constrained economies affected by climate change: nonlinear and stochastic effects. In: Tenreiro Machado JA, et al (eds) Mathematical topics on modelling complex systems, nonlinear physical science
25. Narita D, Tol RS, Anthoff D (2010) Economic costs of extratropical storms under climate change: an application of FUND. J Environ Plann Manag 53(3):371–384
26. von Neumann J (1937) Über ein Ökonomisches Gleichungssystem und eine Verallgemeinerung des Brouwerschen. Ergebnisse eines Mathematische Kolloquiums 8
27. von Neumann J (1945) A model of general equilibrium. Review of Economic Studies 13:1–9
28. Nordhaus WD (2008) A Question of Balance. Yale University Press, New Haven & London
29. Pfeifer S, Rechid D, Reuter M, Viktor E, Jacob D (2019) 1.5°, 2°, and 3° global warming: visualizing European regions affected by multiple changes. Reg Environ Change 19(6):1777–1786
30. Preuschmann S, Hänsler A, Kotova L, Dürk N, Eibner W, Waidhofer C, Haselberger C, Jacob D (2017) The IMPACT2C web-atlas—Conception, organization and aim of a web-based climate service product. Clim Serv 7:115–125
31. Rabinovich MI, Huerta R, Varona P, Afraimovich VS (2008) Transient cognitive dynamics, metastability, and decision making. PLoS Comput Biol 4(5):e1000072

32. Teichmann C, Bülow K, Otto J, Pfeifer S, Rechid D, Sieck K, Jacob D (2018) Avoiding extremes: benefits of staying below +1.5 °C compared to +2.0 °C and +3.0 °C global warming. Atmosphere 9:115
33. Volchenkov D (2016) Survival under Uncertainty. An Introduction to Probability Models of Social Structure and Evolution. Springer International Publishing
34. Volchenkov D, Helbach J, Tscherepanow M, Küheel S (2013) Exploration-exploitation trade-off in a treasure hunting game. Electron Notes Theor Comput Sci 299:101–121
35. Weber M, Barth V, Hasselmann K (2005) A multi-actor dynamic integrated assessment model (MADIAM) of induced technological change and sustainable economic growth. Ecol Econ 54(2):306–327
36. Weber T, Haensler A, Rechid D, Pfeifer S, Eggert B, Jacob D (2018) Analyzing regional climate change in Africa in a 1.5, 2, and 3°C global warming world. Earth's Future 6(4):643–655
37. Weitzman ML (2012) GHG targets as insurance against catastrophic climate damages. J Publ Econ Theory 14(2):221–244

A Regulated Market Under Sanctions. On Tail Dependence Between Oil, Gold, and Tehran Stock Exchange Index

Abootaleb Shirvani and Dimitri Volchenkov

Abstract We demonstrate that the tail dependence should always be taken into account as a proxy for systematic risk of loss for investments. We provide the clear statistical evidence of that the structure of investment portfolios on a regulated market should be adjusted to the price of gold. Our finding suggests that the active bartering of oil for goods would prevent collapsing the national market facing the international sanctions.

Keywords Regulated markets · Tehran stock exchange · Financial econometrics · Tail dependence

1 Introduction

Since 1979, the UN Security Council and the United States regularly passed a number of resolutions imposing economic sanctions on Iran regarding supporting for Iran's nuclear activities [1]. Over the years, sanctions have taken a serious toll on Iran's economy and people, as well as resulted in the increasing government control over the forces of supply and demand and prices in Iran.

Oil price changes have a significant effect on economy since oil prices directly affect the prices of goods and services made with petroleum products. Increases in oil prices can depress the supply of other goods, increasing inflation and reducing economic growth [2]. Being an energy superpower, Iran has an estimated 158 bn barrels of proven oil reserves, representing almost 10% of the world's crude reserves and 13% of reserves held by the Organization of Petroleum Exporting Countries [3]. The Petroleum industry in Iran accounted for 60% of total government revenues and

A. Shirvani · D. Volchenkov (✉)
Department of Mathematics and Statistics, 1108 Memorial Circle, Lubbock, TX 79409, USA
e-mail: dimitri.volchenkov@ttu.edu

A. Shirvani
e-mail: abootaleb.shirvani@ttu.edu

80% of the total annual value of both exports and foreign currency earnings in 2009 [4]. The oil price volatility should strongly affect the Tehran Stock Exchange index (TSE).

Gold is one of the basic assets ensuring stability of national currency. In an economic downturn, people tend more towards gold as a safe asset while reducing their investments that leads to deeper economic downturn [5]. Iran's gold bar and coin sales tripled to 15.2 tons in the second quarter of 2018, the highest in four years as announced by the World Gold Council. According to Bloomberg, Iran's demand for gold bars and coins may remain strong for the rest of the year and even increase as the US reimposes sanctions [6]. The gold price volatility should strongly affect the Tehran Stock Exchange index (TSE).

The main objective of this paper is to estimate stochastic dependence between the daily log-returns for oil and gold and TSE by applying the copula method or constructing non-Gaussian multivariate distributions and understanding the relationships among multivariate data. We also assess the degree of dependence between extreme events on the national oil and gold markets and TSE. Although the linear correlation analysis shows positive correlations between the quantities of interest, the copula method indicates that the degree of dependence between oil and TSE is weak while there might be a significant left tail dependence between TSE and gold that can be thought of as a proxy for systemic risk of default.

We use daily data for the closing Euro Brent Crude-oil price (in US dollars per share) and closing gold historical spot price (in US dollars per ounce) obtained from the Historical London Fix Prices [7]. The daily adjusted closing price for the TSE were taken from the official web site of the Tehran Stock Exchange [8]. The time stamps of TSE prices were converted from the Persian to Western dates. The daily log-returns of data time series (for gold, oil, and TSE) for the period from 28/02/2005 to 14/11/2018 are plotted in Fig. 1.

In Sect. 2, we provide a literature review on the tail dependence between oil, gold, and stock market indices in Iran and worldwide. In Sect. 3, we give a brief introduction to the copula method. Our main contribution is explained in Sect. 4. We apply the tail copula method to investigate the dependence of joint extreme events for the gold and oil prices and TSE index. The standard ARMA–GARCH model is used to estimate the marginal distributions and to filter out the serial dependence and volatility clustering in the data. Several copula models are implemented to the standardized residuals of each series. We estimate and model the tail dependence for the gold and oil prices and TSE index empirically and theoretically. We then conclude in the last section.

2 Literature Review

The impact of crude oil and gold prices on financial markets has been broadly discussed in the literature. Hamilton [9] had shown that the oil price has an essential impact on the economy and its volatility leads to a stock market price change. Kling

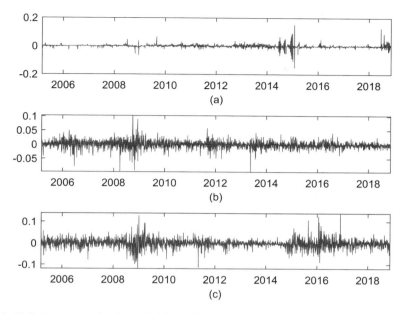

Fig. 1 Daily log-returns for the period from 28/02/2005 to 14/11/2018 for the (**a**) Tehran Stock Exchange; (**b**) Gold; (**c**) Crude Oil price

[10] applied the Vector Autoregressive (VAR) model to assess the impact of oil price movement on the S&P 500 Index and five US industries. Huang et al. [11] considered the relationship between log-return of oil and US stock market, by using VAR model and demonstrated that the oil futures might affect the stock returns of oil companies. Sadorsky [12] used the unrestricted VAR model with generalized autoregressive conditional heteroskedasticity (GARCH) model to show the negative impact of increasing oil prices on the US stock market.

Cai et al. [13] observed that the GDP per capita and the inflation rate have a substantial effect on the gold price volatility. Baryshevsky [14] observed the high inverse correlation between real stock returns and the ten years average rate of gold. Nandha and Faff [15] reported the strong correlation between oil prices and the stock market in oil exporting countries.

Miller and Ratti [16] found that after 1994 the long-term response of market indices to the oil price shocks was negative. Baur and Mcdermott [17] studied the impact of gold price on the financial market during 1979–2009 and observed that gold acts as a hedge in the stock market of the US and most European countries. The empirical findings of Batten et al. [18] was that volatility of the gold return has a substantial impact on financial market returns.

Recently, several authors have applied time series models and the copula method to measure co-movements in the tails of the multivariate distributions. Bharn and Nikolovann [19] considered the relationship between oil price and stock markets in Russia by applying the exponential GARCH model. Fillis et al. [20] studied the time-varying dependence between oil price and stock market in different countries during

1997–2009 by applying the DCC-MGARCH model. Arouri et al. [21] measured the degree of dependence between the oil price shocks and the Gulf Cooperation Council (GCC) stock market by the VAR-GARCH model. Jäschke et al. [23] modeled large co-movements of the commodity returns by applying the copula method to extreme events on the energy market.

Applying the ARJI-GARCH model, Chang [22] observed that the tail dependence between crude oil spot and the future market is time-varying and asymmetric. With the use of Archimedean copulas, Nguyen and Bhatti [24] found no significant evidence of tail dependence between the stock market indices and oil price changes in China and Vietnam. Aloui et al. [25] implemented three Archimedean copulas to measure the significant impact of crude oil price changes on the stock market indices of Gulf Cooperation Council economies.

Mensi et al. [26] studied volatility transmission across the gold, oil, and equity markets and demonstrated the effect the gold and oil price variations have on the S&P 500 index. The copula method had been used for exploring the tail dependence between the financial and credit default swap markets by Silva et al. [27]. Arouri et al. [28] observed the significant impact of gold price volatility on the Chinese stock market in 2004–2011 by applying the VAR-GARCH model. Using the quantile regression method, Zhu et al. [29] found the strong dependence at the upper and lower tails between the crude oil prices and the Asia-pacific stock market index for 2000–2016. Siami-Namini and Hudson [30] examined the volatility spillover from the returns on crude oil to the commodities returns by using the Autoregressive (AR) model with the Exponential GARCH model, in the period from Jan 2006 to Nov 2015. Trabelsi [31] studied the asymmetric tail dependence between international oil market and the Saudi Arabia sector indices in 2007–2016. Hamma et al. [32] applied the copula method and ARMA–GARCH–GDD model to analyze the dependence between stock market indices of Tunisia and Egypt and the crude oil price in 1998–2013.

Few scholars have investigated the dependence between the oil and gold prices and the stock market index in Iran. Foster and Kharazi [33] found no significant correlation between the variations of oil price and TSE in 1997–2002. Applying the ARMA-copula method, Najafabadi et al. [34] observed that the gold and oil price changes might weakly influence TSE. Shams and Zarshenas [35] agreed on that there is no significant evidence of dependence between the oil and gold price variations and the TSE index.

3 Methods

We investigate time series of the daily log-returns,

$$r_t = \log\left(\frac{S_t}{S_{t-1}}\right),$$

$$(1)$$

for the closing prices of an asset S_t.

3.1 Fitting the Time Series by the ARMA–GARCH Model

The ARMA–GARCH model [36, 37] is a standard tool for modeling the conditional mean and volatility of time series. The GARCH model captures several important characteristics of financial time series, including the heavy tail distribution of returns and volatility clustering. The ARMA–GARCH model filters out the linear and non-linear temporal dependence in bi-variate time series,

$$
\begin{aligned}
r_t &= \mu + \sum_{t=i}^{p} \varphi_i \, (r_{t-i} - \mu) + \sum_{j=1}^{q} \theta_j \, a_{t-j} + a_t, \\
a_t &= \varepsilon_t \sigma_t, \quad \varepsilon_t \sim \text{iid}, \\
\sigma_t^2 &= \gamma + \sum_{m=1}^{k} \beta_m \, \sigma_{t-m}^2 + \sum_{n=1}^{l} \alpha_n \, a_{t-n}^2.
\end{aligned}
\tag{2}
$$

where r_t is the assets return, μ in constant term, p and q are the lag orders of ARMA model, k and l are the lag orders of GACH model, $\sigma_t = var \, (r_t \mid F_{t-1})$ is the conditional variance during the period t, F_{t-1} denotes the information set consisting of all linear functions of the past returns available during the time period $t - 1$, ε_t is the standardized residual during the time period t, which are *iid* with zero mean and unit variance; a_t is referred to the shock of return during the period t; $\gamma \geq 0$ is a constant term; $\theta_j \, \{j = 1, 2, \ldots, q\}$, $\varphi_i \, \{i = 1, 2, \ldots, p\}$, $\alpha_n \geq 0 \, \{n = 1, 2, \ldots, l\}$ and $\beta_m \geq 0 \, \{m = 1, 2, \ldots, k\}$ are the parameters of the model estimated from the data.

3.2 Tail Copula Method

A copula is a multivariate probability distribution for which the marginal-probability distribution of each variable is uniform [38]. Application of the copula method to description of the dependence between random variables in finance is relatively new.

In line with Sklar's Theorem [38], every cumulative bivariate distribution F with marginal distributions F_1 and F_2 can be written as

$$
F (x_1, x_2) = C (F_1 (x_1), F_2 (x_2)),
\tag{3}
$$

for some copula C, which is uniquely determined on the interval $[0, 1]^2$.

Conversely, any copula C may be used to design a joint bivariate distribution F from any pair of univariate distributions F_1 and F_2, viz.,

$$
C (u, v) = F \left(F_1^{-1} (u), F_2^{-1} (v) \right),
$$

where F_1^{-1} and F_2^{-1} are the quantile functions of the respective marginal distributions. If we have a random vector $X = (X_1, X_2)$, the copula for their joint distribution is

$$C(u, v) = P(U \le u, V \le v) = F\left(F_1^{-1}(u), F_2^{-1}(v)\right), \qquad (4)$$

for all $u, v \in [0, 1]$.

We also use the survival copula $\bar{C}$ that links the joint survival function $\bar{F}(x) = 1 - F(x)$ to the univariate marginal distribution, viz.,

$$\bar{C}(u, v) = \bar{C}\left(\bar{F}_1(u), \bar{F}_2(v)\right) = \Pr(X_1 \ge u, X_2 \ge v). \qquad (5)$$

Following [39], we can write survival copula function as

$$\bar{C}(u, v) = \Pr(U \ge 1 - u, V \ge 1 - v) = \bar{F}\left(F_1^{-1}(1 - u), F_2^{-1}(1 - v)\right). \qquad (6)$$

3.3 Rank Correlation and Tail Dependence Coefficients

The Kendall and Spearman rank correlation coefficients of two variables, X_1 and X_2, with the copula $C(u, v)$ are given by

$$\tau(X_1, X_2) = 4 \int_0^1 \int_0^1 C(u, v) \, dC(u, v) - 1, \qquad (7)$$

and

$$\rho_S(X_1, X_2) = 12 \int_0^1 \int_0^1 C(u, v) \, du \, dv - 3, \qquad (8)$$

respectively.

Let the distribution of X_1 and X_2 denote by F_1 and F_2. The following relation exists between Spearman's rank and linear correlation coefficients:

$$\rho_s(X_1, X_2) = C(F_1(x_1), F_2(x_2)), \qquad (9)$$

where $(u, v) = (F_1(x), F_2(x))$.

The relations between Kendall's Tau and Spearman's rank correlation coefficients and the coefficient of linear correlations in the Gaussian and Student's t-copulas are

$$Corr(X_1, X_2) = \sin\left(\frac{\pi}{2}\tau\right), \qquad (10)$$

$$Corr(X_1, X_2) = \sin\left(\frac{\pi}{6}\rho_s\right). \qquad (11)$$

Both ρ_s and τ may be considered as a measure of the degree of monotone dependence between random variables, whereas the linear correlation coefficient measures the degree of linear dependence only. Since τ and ρ_s measure the dependence in centered data, they are often insufficient to estimate and describe the dependence structure of extreme events. Hence, according to Embrechts et al. [40], it is significantly better to use the tail copula method than the linear correlation coefficient to characterize the dependence of extreme events. In their opinion, one should choose a model for the dependence structure that reflects more detailed knowledge of the value at risk, and the tail copula method is an excellent tool for managing risks against concurrent events.

The standard way to assess tail dependence is to look at the lower and upper tail coefficients denoted by λ_l and λ_u, respectively, where λ_u quantifies the probability to observe a large X_1 value given the large value of X_2. Similarly, λ_l is a measure that quantifies the probability to observe a small X_1 value, assuming that the value of X_2 is small. Let $X_i \sim F_{X_i}$ and the probability $\alpha \in (0, 1)$ then the upper tail coefficient is:

$$\lambda_u (X_1, X_2) = \lim_{\alpha \to 1} \Pr\left(X_1 \geq F_1^{-1}(\alpha) \mid X_2 \geq F_2^{-1}(\alpha)\right), \qquad (12)$$

and similarly

$$\lambda_l (X_1, X_2) = \lim_{\alpha \to 0} \Pr\left(X_1 \leq F_1^{-1}(\alpha) \mid X_2 \leq F_2^{-1}(\alpha)\right). \qquad (13)$$

On the one hand should $\lambda_l (X_1, X_2) = 0$ ($\lambda_u (X_1, X_2) = 0$) then X_1 and X_2 are said to be lower (upper) asymptotically independent. On the other hand should $\lambda_l (X_1, X_2) > 0$ ($\lambda_u (X_1, X_2) > 0$), then the small (large) events tend to happen coherently, and X_1 and X_2 are lower (upper) tail dependent.

According to Jäschke [23], the Value-at-Risk (VaR) is closely related to the concept of tail dependence. Tail dependencies can be considered as the degree of likelihood of an asset return falling below its VaR at the certain level α when the other asset returns have fallen below its VaR at the same level. In general, λ_l and λ_u (like the other scalar quantities) describe a certain level of dependence in tails. However, in our analysis, we need to describe the general framework of tail dependence for bi-variate distributions.

In the general framework of tail copulas, the dependence structure of extreme events in bi-variate distributions, independently of their marginals, is represented by tail dependence (see Schimtz and Stadtmüller [41] for more details).

The lower and upper tail dependence associated with X_1 and X_2 are

$$\Lambda_L (X_1, X_2) = \lim_{t \to \infty} t\, C\left(\frac{x_1}{t}, \frac{x_2}{t}\right), \qquad (14)$$

$$\Lambda_U (X_1, X_2) = \lim_{t \to \infty} t\, \bar{C}\left(\frac{x_1}{t}, \frac{x_2}{t}\right) \qquad (15)$$

provided the above limits exist everywhere on $\mathbb{R}_+^2 := [0, \infty)^2$. According to Schimtz and Stadtmüller [41], the estimation of tail dependence is a nontrivial task, particularly for a non-standard distribution that is why we consider the tail coefficient as a measure of the tail dependence. The tail coefficient is a specific case of tail dependence, and we have $\lambda_l = \Lambda_L (1, 1)$ and $\lambda_u = \Lambda_U (1, 1)$.

4 Results

4.1 Model Selection

First, we perform the Ljung-Box Q-test [42] to examine the presence of autocorrelation in log-returns and the presence of heteroskedasticity in squared log-return of each data set for lags 5 and 10. The p-values ($p < 0.01$) indicate that the log-return of data (on gold, oil, and TSE) are autocorrelated in each time series. There are significant heteroskedasticity effects, since all p-values are less than 0.01. The hypothesis of independent and identically distributed data is therefore rejected.

Second, to investigate the multiple co-integration relationships among the gold, crude oil, and TSE time series, we apply the Engle-Granger co-integration test [36, 37]. The obtained p-values ($p_1 = 0.0102$, $p_2 = 0.0000$) reject the null hypothesis of no co-integration among the time series. The test result indicates that there is a long-run relationship among the variables, and they share a common stochastic drift.

Third, the linear and nonlinear temporal dependencies in bi-variate time series should be filtered out by applying the ARMA–GARCH model [36, 37]. We tried the ARMA-GRACH model with various lags and different distributions to select the optimal model for each time series. Namely, we have tested the normal, Student's t, generalized error, skewed Student's t, and generalized hyperbolic distributions.

To examine the presence of long memory dependence in each time series, we perform the fractional ARMA(1, d, 1)–GARCH(1, 1) test. The results of all tests are given in Table 1.

The test p-values for gold and oil log-returns ($p \geq 0.05$)) demonstrate *no long memory dependence* in oil and gold time series. However, the very small p-value for TSE log-return ($p = 0.000$) shows that there is a long memory dependence in the TSE time series. The fractional exponent ($d \simeq 0.142$) of the shift operator $(1 - B)$ confirms the presence of long memory dependence in the TSE log-returns.

Fourth, we have evaluated all possible ARMA–GARCH models ($p \leq 2, q \leq 2, k \leq 2, l \leq 2$) with the use of the R-Package "*rugarch*" [43] by examining (i) Akaike Information Criterion (AIC); (ii) Bayesian Information Criterion (BIC) [36, 37]; (iii) the statistical significance test of model parameters at the 5%-level; (iv) lack of autocorrelation; (v) lack of heteroskedasticity in standardized residuals of each time series for lags 5 and 10.

Table 1 Long memory test in TSE, Oil, and Gold log-return

Data	FARIMA(1,d,1)–GARCH(1,1)	
	d	p-value
TSE	0.142	0.000
Oil	0.042	0.059
Gold	0.050	0.102

Table 2 Maximum likelihood estimation, standard errors, and estimations of model parameters in FARMA(1, 0)–GARCH(1, 1) with SGHYD for TSE, the ARMA(1, 1)–GARCH(1, 1) model with Student's t-distribution for oil, and the GARCH(1, 1) model with SGHYD for gold

Parameter	TSE log-return		Oil log-return		Gold log-return	
	Estimation	Std. error	Estimation	Std. error	Estimation	Std. error
ARMA–GARCH model						
φ_1	0.1477	0.0273	0.8705	0.2250	–	–
θ_1	–	–	−0.7850	0.2239	–	–
α_1	0.3616	0.0277	0.0348	0.0037	0.0285	0.0022
β_1	0.6374	0.0441	0.9642	0.0021	0.9655	0.0019
d-(arfima)	0.1331	0.0203	–	–		
Distribution						
ν(DF)	0.2500	0.0158	3.3629	0.2743	0.2500	0.0253
ζ	−0.0327	0.0613	–	–	0.0035	0.0187
η	−1.1811	0.1183	–	–	0.3665	0.0657

Fifth, to evaluate the density of standardized residuals (innovations) in each time series, we apply probability integral transform method following [44]. This method is based on the relation between the sequence of densities of the standardized residuals, $p_t(z_t)$, and its integral probability transform,

$$y_t = \int_{-\infty}^{z_t} p_t(u)\, du. \tag{16}$$

We evaluate the densities of the standardized residuals by assessing whether the probability integral transform series, $\{y_t\}_{t=1}^m$, are $iid\ U(0, 1)$. The non-parametric Kolmogorov-Smirnov test is the easiest way to check the uniformity of y_t's [45].

By comparing all criteria and evaluating the densities of the standardized residuals, the obtained time series models are the FARIMA $(1, 0.14, 0)$–GARCH $(1, 1)$ model with the Standardized Generalized Hyperbolic Distribution (SGHYD), the ARMA $(1, 1)$–GARCH $(1, 1)$ model with Student's t-distribution with $v = 3.363$ degree of freedom, and the GARCH$(1, 1)$ model with SGHYD, for TSE, gold, and oil time series, respectively. The estimated parameters of each model obtained by using the maximum likelihood methods are given in Table 2.

To study autocorrelation and conditional heteroskedasticity in each residual time series, we have shown the QQ plots, and correlograms for z_t and z_t^2, for each standardized residual series in Fig. 2. The apparent linearity of the QQ-plots shows that the corresponding distributions are well-fitted.

The parameters for each model were estimated by the maximum-likelihood method and all parameters are found significant at the 5%-level. Since the p-values in the Ljung-Box Q-tests are less than 0.01, there is no significant autocorrelation and heteroskedasticity of the standardized residuals at the 5%-level for each time series.

A. Shirvani and D. Volchenkov

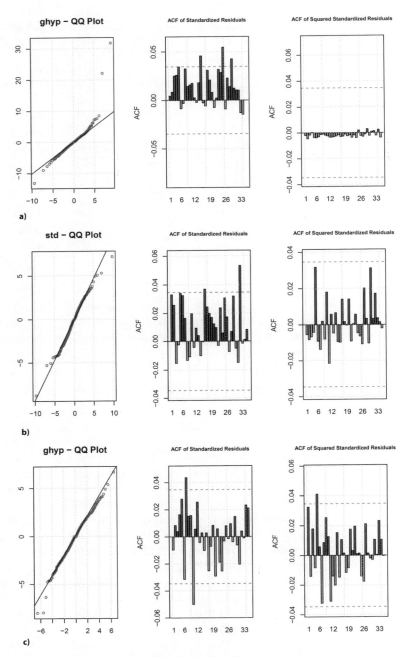

Fig. 2 QQ plot and correlogram of ACF of standardized residuals and ACF of squared standardized residuals for (**a**) TSE log-return, (**b**) Oil log-return, (**c**) Gold log-return

For evaluating the forecast densities of each model, first we calculate the $\{y_t\}_{t=1}^m$ for each standardized residual sets associated with gold, oil and TSE. Then, we test that y_t's are $iid\ U(0, 1)$ by the Kolmogorov-Smirnov and Sarkadi-Kosik tests [46] with the use of the R-package "*uniftest*" [47]. The p-values ($p \geq 0.05$) indicate that y_t's are $iid\ U(0, 1)$ at the 5%-level.

We conclude that the obtained distributions are well fitted.

Therefore, the optimal ARMA–GARCH model for the oil log-returns is

$$
\begin{aligned}
O_t &= 0.000343 + 0.870548\, O_{t-1} - 0.784959\, a_{t-1} + a_t, \\
a_t &= o_t\, \sigma_t, \quad o_t\ iid\ \sim\ t_{(v=3.363)}, \\
\sigma_t^2 &= 0.034843\, o_{t-1}^2 + 0.954157\, \sigma_{t-1}^2\,.
\end{aligned}
\tag{17}
$$

The best ARMA–GARCH model for the gold log-returns is

$$
\begin{aligned}
G_t &= 0.000175 + a_t, \\
a_t &= g_t\, \sigma_t, \quad g_t\ iid\ \sim\ SGHYD\,(0.00348, 0.2500), \\
\sigma_t^2 &= 0.028526\, g_{t-1}^2 + 0.945546\, \sigma_{t-1}^2\,.
\end{aligned}
\tag{18}
$$

Finally, the optimal model for the log-returns for TSE is

$$
\begin{aligned}
(1 - B)^{0.14} S_t &= 0.000251 + 0.147724\, S_{t-1} + a_t, \\
a_t &= s_t\, \sigma_t, \quad s_t\ iid\ \sim\ SGHYD\,(-0.032714, 0.2501), \\
\sigma_t^2 &= 0.361576\, s_{t-1}^2 + 0.627424\, \sigma_{t-1}^2\,.
\end{aligned}
\tag{19}
$$

where B is back-shift operator. The *rugarch* [43] package performs GHYD parameter estimations using the (η, ζ) parametrization (SGHYD), after which a series of steps transform those parameters into the $(\mu, \alpha, \beta, \delta)$ while at the same time including the necessary recursive substitution of parameters in order to standardize the resulting distribution.

4.2 Tail Dependence Measure

4.2.1 Selection of the Copula Model

In this section, we model the dependence structure between the log-returns of oil, gold, and TSE denoted by O_t, G_t, and S_t, respectively, by the copula method. We have already shown that the O_t, G_t and S_t are time-varying dependence series. We applied the time series model to filter out time-varying dependence and obtain standardized residuals for gold, oil, and TSE, denoted by g_t, o_t, and s_t, respectively. We work with the *iid* standardized residuals for each time series.

Deheuvels et al. [48] introduced the empirical copula based on the rank correlation. An initial approach to calculate the empirical copula function is to estimate the distribution function of residuals based on the empirical distribution function F_X,

$$F_X(x) = \frac{1}{T+1} \sum_{i=1}^{T} \mathbb{1}\,(X_i \leq x), \tag{20}$$

where $\mathbb{1}(A)$ denotes the indicator function of the set A. The quantities $F_X(x_i)$ and $F_Y(y_i)$ as given by (20) are the ranks of X_i and Y_i normalized by $T + 1$.

Residual sets o_t, g_t, and s_t are transformed to the rank-based variables by

$$u_t = \frac{rank\,(o_t)}{T+1}, \quad v_t = \frac{rank\,(g_t)}{T+1}, \quad w_t = \frac{rank\,(s_t)}{T+1}. \tag{21}$$

The domain of empirical copula, $(u, v) \in [0, 1]^2$ [49], is formed by all normalized ranks, $\left\{ \frac{1}{T+1}, \frac{2}{T+1}, \ldots, \frac{T}{T+1} \right\}$, for each residual set. The best sample-based empirical copula is then defined by

$$C_u(u, v) = \frac{1}{T} \sum_{t=1}^{T} \mathbb{1}\,(U_t < u, V_t < v), \tag{22}$$

and the empirical survival copula is

$$\bar{C}_u(u, v) = \frac{1}{T} \sum_{t=1}^{T} \mathbb{1}\,(U_t \geq u, V_t \geq v). \tag{23}$$

Before using the obtained copulas, we have estimated the Spearman ρ and Kendall τ rank correlation coefficients from the normalized ranks (see Table 3). The obtained coefficients show that the dependence between oil and TSE is very weak, but the correlation between gold and TSE is significant. The dependence between gold, oil, and TSE was also assessed by applying the multivariate independent test [50]. The obtained p-value ($p = 0.021$) at the 5% level rejects the hypothesis of independence of these variables. We conclude that the Spearman and Kendall's correlations can not measure the entire dependence between the quantities of interest.

Now we determine the optimal copula model to describe the dependence structure of the joint distributions. We consider both theoretical (parametric) C and empirical (non-parametric) C_u methods. We fit several copula families [Independent, Gaussian, Student's t, Clayton, Gumbel, Frank and Joe] on standardized residuals. The unknown copula parameter θ is obtained by the *inverse τ method*, i.e., by solving the equation (11) for the value $\hat{\tau}$ estimated from the data with the use of the R-

Table 3 Correlation between ranks of standardized residual of gold, oil and TSE

Method	Kendall		Spearman	
	Gold	Oil	Gold	Oil
TSE	0.111	0.017	0.162	0.026

Table 4 The copula parameters and goodness-of-test results for the different copula families

Copula	(a) Oil and TSE				(b) Gold and TSE			
	$\hat{\theta}$	AIC	BIC	p-value	$\hat{\theta}$	AIC	BIC	p-value
Independent				0.79				0.00
Gaussian	0.037	−2.249	3.822	0.83	0.174	−82.875	−76.803	0.02
t	0.034	6.544	18.688	0.04	0.173	−66.765	−54.622	0.00
Clayton	0.028	−0.451	5.621	0.63	0.161	−44.561	−38.489	0.40
Gumbel	1.016	−1.285	4.787	0.75	1.080	−50.448	−44.377	0.07
Frank	0.151	−0.053	6.019	0.34	1.003	−80.859	−74.788	0.00
Joe	1.019	−0.666	5.405	0.76	1.080	−19.940	−13.868	0.63

package 'VineCopula' [51]. The parameter θ controls the strength of dependence in each copula family; its values are collected in Table 4. The significance of each fitted copula is examined by the goodness of fit test with the use of Cramer-Von-Mises statistics [52]:

$$S_t = \sum_{t=1}^{T} \left(C_t \left(u_t, v_t \right) - C_{\hat{\theta}_t} \left(u_t, v_t \right) \right)^2. \tag{24}$$

The corresponding p-value is obtained by the bootstrapping method (see [53] for details). We have evaluated each copula family with respect to the following three criteria: (i) the high p-value of the goodness of fit test; (ii) the AIC; and (iii) the BIC values. The obtained values are given in Table 4.

We conclude that the Student's t copula is not appropriate for modeling the dependence between oil and TSE (as p-value ≤ 0.05). The small goodness-of-fit p-values for gold and TSE also reject the Independent, Frank, and t copula families at the 5% level. For any other candidates, the p-value can not reject the null hypothesis.

The Gaussian copula is a candidate for the optimal model describing the dependence between oil and TSE, with the smallest AIC and BIC values, and the highest p-value among all other candidates.

Due to zero p-value, the Frank copula is not a good fit for the description of dependence between gold and TSE although the values of AIC (-80.859) and BIC (-74.788) are the smallest ones. A copula family with the non-trivial tail coefficients and higher p-value goes over other families in explaining the entire dependence between gold and TSE. Therefore, we have selected the Joe copula as a proper candidate because of its higher p-value (0.63) and the smaller AIC and BIC values over the Clayton copula even though the lower tail dependence coefficient in Joe copula is zero.

4.2.2 Lower and Upper Tail Coefficients for Oil, Gold, and TSE

Jäschke et al. have shown that the goodness-of-fit test alone does not necessarily provide an appropriate model for tail dependence, because it is based on minimizing the distance between the observed ranks model and parametric model over the whole support of the distribution [23]. They suggested applying the tail copulas concept for capturing dependence in the tail of distribution to improve the effectiveness of fitted copula. The non-parametric estimators are defined up to a scaling factor $k = 1, 2, \ldots, T$ chosen by a statistician [41], for the lower tail copula,

$$\hat{\Lambda}_L(x, y) = \frac{T}{k} C_u \left(\frac{kx}{T}, \frac{ky}{T} \right) = \frac{1}{k} \sum_{t=1}^{T} \mathbb{1} \left(u_t \leq \frac{kx}{T+1}, v_t \leq \frac{ky}{T+1} \right), \quad (25)$$

and for the upper tail copula,

$$\hat{\Lambda}_U(x, y) = \frac{T}{k} \bar{C}_u \left(\frac{kx}{T}, \frac{ky}{T} \right) = \frac{1}{k} \sum_{t=1}^{T} \mathbb{1} \left(u_t > \frac{T-kx}{T+1}, v_t > \frac{T-ky}{T+1} \right), \quad (26)$$

respectively. Based on the above estimators, they found that

$$\hat{\lambda}_l = \hat{\Lambda}_{L,T}(1, 1), \quad \hat{\lambda}_u = \hat{\Lambda}_{U,T}(1, 1) \quad (27)$$

are the appropriate non-parametric estimators for the upper and lower tail dependence coefficients.

We estimate the lower and upper tail coefficients by using the Eq. (27). The values of coefficient estimators for tail dependence between oil and TSE, gold and TSE are given in Table 5. In the preceding section, our results based on the highest p-value, the smallest AIC and BIC values have suggested that the Gaussian copula is the optimal one for describing the dependence between oil and TSE. However, the zero values of estimators given in Table 5 indicate that there is no significant tail dependence between oil and TSE, in line with the small values of the linear correlation coefficients given in Table 3. Therefore, we conclude that the independent copula is the optimal model for describing the dependence between oil and TSE on the Iranian market.

Concerning the tail dependence between gold and TSE, the results of the previous subsection show that the Joe copula is optimal for describing the dependence structure between gold and TSE. The non-parametric estimators for the upper and lower tail dependence between gold and TSE given in Table 5. By comparing the empirical coefficients with the parametric tail coefficient of Joe copula, we conclude that the Joe copula, which has the nontrivial upper tail and zero lower tail dependence, can not well enough explain the risk of extreme events in the tails.

From Table 5, we see that the Gumbel copula goes over other copula families in explaining the dependence structure between gold and TSE due to the small AIC and

Table 5 Tail dependence coefficient estimators for tail dependence between oil and TSE, gold and TSE for the different copula families

Copula	Oil and TSE		Gold and TSE	
	Lower	Upper	Lower	Upper
Independent	0.0000	0.0000	0.0000	0.0000
Gaussian	0.0000	0.0000	0.0000	0.0000
Clayton	0.0000	0.0000	0.0434	0.0000
Gumbel	0.0000	0.0216	0.0000	0.1000
Frank	0.0000	0.0000	0.0000	0.0000
Joe	0.0000	0.0250	0.0000	0.1005
Empirical	0.0000	0.0000	0.0680	0.0000

BIC values. However, comparing the empirical coefficients with the parametric tail coefficients of Gumbel copula, we see that the former is inverse with respect to the Gumbel ones. Therefore, we should reject the Gumbel copula as well.

We conclude that the Clayton copula is the optimal model for describing the dependence structure between gold and TSE on the Iranian market, because the lower tail coefficient is close to empirical while the upper tail is zero. The Clayton copula has the second highest p-value among the other candidates, and its AIC and BIC values are small comparing to those for other copulas (see Table 5).

5 Discussions and Conclusions

Tails matter! In our work, we convincingly demonstrate that the standard goodness measures for fitting copulas to data, such as the p-value, AIC/ BIC values, are not the reliable indicators of goodness-of-fit. The tail dependence should always be taken into account as a proxy for systemic risk by risk managers.

Although the small AIC/BIC values and the goodness-of-fit test assume that the Gaussian copula is the good one for representing the dependence structure between oil and TSE on the Iranian market, the absence of tail dependence and low correlation between these assets disclose that the independent copula is the optimal one for fitting the data. Again, the small AIC/BIC values, the goodness-of-fit test, and the high p-value suggest that the Joe copula would be a good data model describing the relation between gold and TSE on the Iranian market. However, our finding reveals that the Joe copula, which has only an upper tail and no lower tail dependence, fails to describe the tail dependence of gold and TSE properly. The empirical tail coefficients suggest that the Clayton copula might be a suitable model fitting the data on gold and TSE well.

Watch the gold price! Our finding has the important implications for risk managers and investors, because of it can help them to adjust the structure of their investment portfolios once the gold price changes. Risk managers should reduce the ratio of their investments into TSE whenever the gold price is decreasing.

Barter when facing sanctions! Our finding has also the important implications for policy maker of countries faced international sanctions. The lack of dependence between the TSE index and the oil price generating the major foreign currency revenue suggests that oil dollars almost do not influence the national market of Iran.

Faced with the international sanctions, Iran was turning to barter by offering gold bullion in overseas vaults or tankerloads of oil, in return for food as the financial sanctions had hurt its ability to import basic staples [54]. Those sanctions were not banning companies from selling food to Iran, but the transactions with banks were very difficult. Unable to bring in U.S. dollars and euros ahead of the new U.S. sanctions, Iran is open to accepting agricultural products and medical equipment in exchange for its crude oil [55]. A scheme to barter Iranian oil for European goods through Russia, which would then refine it and sell it to Europe as part of mechanism to bypass American sanctions on the Islamic Republic was announced by Europe, China, and Russia along the sidelines of the United Nations General Assembly [56].

In the future, we plan to implement the copula method by using rolling windows for modeling the time-varying dependence between gold, oil and TSE in a multivariate framework.

References

1. Timothy Alexander Guzman (10 April 2013). New economic sanctions on Iran, Washington's Regime Change Strategy. Global Research. Retrieved 5 May 2013
2. Blanchard O, Gali J (2007) The macroeconomic effects of oil shocks: why are the 2000s so different from 1970s. NBER Working Paper No. 13368
3. Financial Tribune, Iran's Proven Oil Reserves to Rise by 10 Percent, Thursday March 09 2019
4. Press.TV Iran oil exports top 844 mn barrels Wed Jun 16, 2010 4:59PM
5. Pirbasti NF, Tajeddini M (2016) Factors affecting on the price of gold on global markets and its impact on the price of gold in Iran market (incorporation of dynamic system pattern and econometric). Modern Appl Sci 10(3); Published by Canadian Center of Science and Education (2016)
6. Financial Tribune, Economy, Business and Markets August 04, 2018 19:51, Gold Demand at Four-Year High
7. Historical London Fix Prices
8. The current information about Tehran Stock Exchange is available
9. Hamilton JD (1983) Oil and the macroeconomy since World War II. J Polit Econ 91(2):228–248
10. Kling JL (1985) Oil price shocks and stock market behavior. J Portf Manag 12:34–39
11. Huang R, Masulis R, Stoll H (1996) Energy shocks and financial markets. J Futur Mark 16(1):1–27
12. Sadorsky P (1999) Oil price shocks and stock market activity. Energy Econ 21:449–469
13. Cai J, Cheung YL, Wong MC (2011) What moves the gold market? J Future Markets 21:257–278
14. Baryshevsky DV (2004) The interrelation of the long-term gold yield with the yields of another asset classes
15. Nandha M, Faff R (2008) Does oil move equity prices? A global view. Energy Econ 30:986–997

16. Miller JI, Ratti RA (2009) Crude oil and stock markets: stability, instability, and bubbles. Energy Econ 31(4):559–568
17. Baur DG, Mcdermott TK (2010) Is gold a safe haven? International evidence. J Bank Finance 34(8):1886–1898
18. Batten JA, Ciner C, Lucey BM (2010) The macroeconomic determinants of volatility in precious metals markets. Resour Policy 35(2):65–71
19. Bharn R, Nikolovann B (2010) Global oil prices, oil industry and equity returns: Russian experience. Scottish J Polit Econ 57(2):169–186
20. Filis G, Degiannakis S, Floros C (2011) Dynamic correlation between stock market and oil prices: the case of oil-importing and oil-exporting countries. Int Rev Financ Anal 20(3):152–164
21. Arouri MEH, Lahiani A, Nguyen DK (2011) Return and volatility transmission between world oil prices and stock markets of the GCC countries. Econ Model 28:1815–1825
22. Chang KL (2012) The time-varying and asymmetric dependence between crude oil spot and futures markets: evidence from the mixture copula-based ARJI–GARCH model. Econ Model 29(6):2298–2309
23. Jäschke S, Siburg KF, Stoimenov PA (2012) Modeling dependence of extreme events in energy markets using tail copulas. J. Energy Markets 5(4):63–80
24. Nguyen C, Bhatti MI (2012) Copula model dependency between oil prices and stock markets: evidence from China and Vietnam. J Int Financ Inst Money 22:758–773
25. Aloui R, Hammoudeh S, Nguyen DK (2013) A time-varying copula to oil and stock market dependence: the case of transition economies. Energy Econ 39:208–221
26. Mensi W, Beljid M, Boubaker A, Managi S (2013) Correlations and volatility spillovers across commodity and stock markets: linking energies, food, and gold. Econ Modell 32:15–22
27. Silva PP, Rebelo P, Afonso C (2014) Tail dependence of financial stocks and CDS markets: evidence using copula methods and simulation-based inference. Economics-The Open-Access, Open-Assessment E-Journal 8(39):1–27
28. Arouri MEH, Lahiani A, Nguyen DK (2014) World gold prices and stock returns in China: insights for hedging and diversification strategies. Econ Modell 44:273–282
29. Zhu H, Huang H, Peng C, Yang Y (2016) Extreme dependence between crude oil and stock markets in Asia-Pacific regions: evidence from quantile regression. Economics discussion papers, No. 2016-46, Kiel Institute for the World Economy
30. Siami-Namini S, Hudson D (2017) Volatility spillover between oil prices, US dollar exchange rates and international agricultural commodities prices. In: Presented at 2017 annual meeting, February 2017, Mobile, Alabama 252845, Southern Agricultural Economics Association
31. Trabelsi N (2017) Asymmetric tail dependence between oil price shocks and sectors of Saudi Arabia System. J Econ Asymmetries 16:26–41
32. Hamma W, Ghorbel A, Jarboui A (2018) Copula model dependency between oil prices and stock markets: evidence from Tunisia and Egypt. Am J Finance Account 2:111–150
33. Foster KR, Kharazi A (2008) Contrarian and momentum returns on Iran's Tehran Stock Exchange. J Int Finan Markets Inst Money 18(1):16–30
34. Najafabadi ATP, Qazvini M, Ofoghi R (2012) The impact of oil and gold prices shock on Tehran stock exchange: a copula approach. Iran J Econ Stud 1(2):23–47
35. Shams S, Zarshenas M (2014) Copula approach for modeling oil and gold prices and exchange rate co-movements in Iran. Int J Stat Appl 4(3):172–175
36. Fuller WA (1976) Introduction to statistical time series. Wiley, New York
37. Hamilton JD (1994) Time series analysis. Princeton University Press, Princeton, N.J
38. Sklar A (1959) Fonctions de répartition á n-dimensions et leurs marges. Publ Inst Statist Univ Paris (in French) 8:229–231
39. Nelsen RB (1999) An introduction to copulas. Springer, New York
40. Embrechts P, McNeil AJ, Straumann D (1999) Correlation: pitfalls and alternatives. Risk Mag 5:69–71
41. Schmidt R, Stadtmüller U (2006) Nonparametric estimation of tail dependence. Scand J Stat 32(2):307–335

42. Ljung GM, Box GEP (1978) On a measure of a lack of fit in time series models. Biometrika 65(2):297–303
43. Ghalanos A (2018) rugarch: Univariate GARCH models. R package version 1.4–0
44. Diebold FX, Gunther TA, Tay AS (1998) Evaluating density forecasts, with applications to financial risk management. Int Econ Rev 39:863–883
45. Kolmogorov A (1933) Sulla determinazione empirica di una legge di distribuzione. G Ist Ital Attuari 4:83–91
46. Kosik P, Sarkadi K (1985) A new goodness-of-fit test. In: Proceedings of the 5th Pannonian symposium on mathematical statistics Visegrad, Hungary, vol 20, pp 267–272
47. Melnik M, Pusev R (2015) uniftest: goodness-of-fit tests for the uniform distribution. R package version 1:1
48. Deheuvels P (1979) La fonction de dépendance empirique et ses propriétés. Un test non paramétrique d'indépendance. Acad Roy Belg Bull Cl Sci 65(6):274–292
49. Genest C, Favre A-C (2007) Everything you always wanted to know about copula modeling but were afraid to ask. J Hydrol Eng 12:347–368
50. Genest C, Kojadinovic I, Neslehová J, Yan J (2011) A goodness-of-fit test for bivariate extreme-value copulas. Bernoulli 17(1):253–275
51. Schepsmeier U, Stoeber J, Brechmann EC, Graeler B (2013) Package VineCopula: statistical inference of vine copulas. R-Project CRAN Repository
52. Genest C, Rémillard B, Beaudoin D (2009) Goodness-of-fit tests for copulas: a review and a power study. Insurance Math Econom 44(2):199–213
53. Genest C, Rémillard B (2008) Validity of the parametric bootstrap for goodness-of-fit testing in semiparametric models. Annales de l'Institut Henri Poincaré - Probabilités et Statistique 44(6):1096–1127
54. Reuters World News, February 9, 2012 / 5:59 AM; Iran turns to barter for food as sanctions cripple imports by Valerie Parent, Parisa Hafezi
55. oilprice.com (2018) Iran looks to barter oil as U.S. Sanctions Bite by Tsvetana Paraskova - Jul 05, 2018, 6:00 PM CDT
56. independent.co.uk (2018) Europe and Iran plot oil-for-goods scheme to bypass US sanctions by Borzou Daragahi, Wednesday 26 September 2018 21:58

Dynamics of Water-Constrained Economies Affected by Climate Change: Nonlinear and Stochastic Effects

Dmitry V. Kovalevsky and María Máñez-Costa

Abstract We study the dynamics of water-constrained regional economies affected by adverse climate change that leads to increasing aridity over time, manifested in growing frequency and severity of droughts. As a basis, we use a family of deterministic economic models that include the AK model, the Solow–Swan model, as well as linear and nonlinear versions of the Structural Dynamic Economic Model (SDEM). We then include in the modelling scheme drought damage functions parametrising adverse economic impacts of droughts. The analysis of regional economic scenarios with drought damages gradually growing over time in deterministic model setup provides first insights to projected future impacts of climate change on a water-constrained economy. We then move to a more detailed analysis of the models by introducing the stochasticity in drought damage functions. This allows modelling the regional economy affected by a series of random fluctuations in water availability. Modelling the stochastic economic dynamics allows to assess uncertainty caused by temporal variability in drought conditions.

1 Introduction

In modelling of individual decision making, and even more in modelling of collective decision-making in the groups of interacting individuals, accounting for both nonlinear and stochastic effects is important [1, 2, 26, 31, 32]. Nonlinearity allows describing systems with substantial structural complexity, while stochasticity sheds light on inevitable uncertainty related to any attempt of modelling such systems.

In particular, nonlinear, stochastic, and nonlinear stochastic models play a prominent role in economic modelling [4, 17]. Models of such kind are even more important for describing the dynamics of coupled socio-natural systems. Natural science com-

D. V. Kovalevsky (✉) · M. Máñez-Costa
Climate Service Center Germany (GERICS), Helmholtz-Zentrum Hereon, Fischertwiete 1, 20095 Hamburg, Germany
e-mail: dmitrii.kovalevskii@hereon.de

M. Máñez-Costa
e-mail: maria.manez@hereon.de

ponents embedded in economic models, like exhaustible natural resources, damages
from pollution and impacts of climate change [22, 34] substantially increase the
overall complexity of the modelling, introduce fundamental nonlinearities and, generally, increase the overall uncertainty of coupled models, as compared to uncertainty
inherent to their natural or economic components taken separately.

The focus of the present study is on modelling the dynamics of water-constrained
regional economies affected by droughts, with a particular attention to related nonlinear and stochastic effects. Droughts are climate-related natural hazards. Therefore,
firstly, drought characteristics experience pronounced temporal variability, that calls
for stochastic description of their economic impacts. Secondly, both frequency and
severity of droughts are projected to increase in many regions of the world as a consequence of climate change [9, 18, 30], and this suggests that their economic impacts
will become more severe in the future than they are now. This projected increase of
drought damages should also be included in related models.

In the present study, we address both of these effects (stochastic drought variability
and drought damages increasing over time) with a family of models of regional
economic dynamics. The rest of this book chapter is organised as follows. In Sect. 2
we introduce the members of the model family, and in Sect. 3 we analyse their
solutions. Deterministic drought damages are included in the modelling scheme in
Sect. 4, while stochastic damages are considered in Sect. 5. Section 6 concludes.

2   Linear and Nonlinear Models of Economic Dynamics

We will describe the dynamics of water-constrained economies with a model family
comprised by four different models: the AK model; the Solow–Swan model; linear
and nonlinear versions of SDEM model. In the present section, we describe the fundamentals of these four models. Drought damage functions allowing for the description
of economic impacts of droughts will be included in the modelling scheme later (in
Sect. 4 for the deterministic model setup; in Sect. 5 – for the stochastic model setup).
In all cases, regional economy will be treated, for simplicity, in the approximation
of closed economy. As discussed below, the four members of the model family are
interrelated in a sense that some of them may be considered as particular or limiting
cases of the others.

2.1   The AK Model

The AK model [4, 11, 20, 21] is probably the simplest possible economic growth
model. Despite its simplicity, it can be applicable at different scales (to describe, for
example, national or regional economy).

The AK model is essentially the model of capital stock dynamics, so the concept
of capital is of fundamental importance here. As discussed in e.g. [4], in certain

interpretations of the AK model capital is understood broadly, including physical, human, natural, and social capital. Decomposing the capital into different kinds is also important for assessing the prospects of economic policy implementation, including policies targeted at adaptation to droughts [7, 15, 35].

Below we will present the AK model in terms of capital per unit of effective labour, k, called for brevity the capital below, as a state variable describing the economic system under study.

Suppose that the output of a closed economic system per unit of effective labour, y, referred to as the output below, can be parametrised by a production function dependent on and linear in k:

$$y = Ak, \tag{1}$$

where $A = $ const is the technology parameter. A constant fraction s of the output (s is called the savings rate, $0 < s < 1$) is invested in physical capital. The remaining fraction of the output, equal to $(1 - s)$, is consumed. The capital dynamics is governed by two main processes, acting in opposite directions: (i) capital growth due to investment, (ii) capital depreciation, when the old units of capital gradually go out of use. The depreciation rate is constant and equal to λ_K. Also, the labour is assumed to change with a constant rate λ_L that might be positive (labour growth), zero (constant labour), or negative (labour decline).

In terms of ordinary differential equations (ODE), this model of economic dynamics can be formalized as

$$\dot{k} = [sA - (\lambda_K + \lambda_L)]k. \tag{2}$$

The AK model is a linear economic growth model. The solution of (2) is straightforward and will be briefly discussed in Sect. 3.1.

2.2 The Solow–Swan Model

The Solow–Swan model [4, 28, 29] can be derived from the AK model by replacing the linear production function (1) with a nonlinear function obeying certain conventional conditions (see the cited literature for the details):

$$y = f(k). \tag{3}$$

Specifically, in what follows we will assume the Cobb–Douglas form of the production function broadly used in economic modelling. For the Cobb–Douglas production function, Eq. (3) takes the form

$$f(k) = Ak^\alpha, \qquad 0 < \alpha < 1. \tag{4}$$

Otherwise, a mental model of the economy is the same as was presented in Sect. 2.1. Therefore, replacing in (2) the AK production function (1) with the Cobb–Douglas production function (4), we easily get a nonlinear capital dynamics equation of the form

$$\dot{k} = sAk^{\alpha} - (\lambda_K + \lambda_L)k \tag{5}$$

that will be further examined in Sect. 3.2.

By setting α in (5) to its upper limit from (4), $\alpha \to 1$, we get back to the linear AK model (2).

2.3   The SDEM Model

The model SDEM (the acronym SDEM stands for the Structural Dynamic Economic Model) was initially proposed in [5] and further developed in [6, 12–14]. In SDEM, the economic dynamics are described as driven by a conflict of interest of two powerful aggregate actors, entrepreneurs and wage-earners. In the present study, we adopt the production function in the Cobb–Douglas form (4), changing, for consistency with previous publications on SDEM, the notation for the technology parameter (A is replaced with ν):

$$y(k) = \nu k^{\alpha}, \qquad 0 < \alpha < 1. \tag{6}$$

The economic dynamics are described by two state variables, capital k and average wage of wage-earners w, that obey the following two-dimensional dynamical system:

$$\dot{k} = (1 - \rho_d)\,(\nu k^{\alpha} - w) - (\lambda_K + \lambda_L)\,k, \tag{7a}$$

$$\dot{w} = \lambda_W \left[q\,(\nu k^{\alpha} - (\lambda_K + \lambda_L)\,k) - w \right]. \tag{7b}$$

The state variables k and w are basically expressed in material units, [goods], that might be converted to monetary units by ascribing a certain price to one 'goods' equivalent.[1]

In brief, the narrative behind the dynamic equations of SDEM (7a)–(7b) is as follows. Entrepreneurs 'own' the output of the economy $y(k) = \nu k^{\alpha}$ and, after paying the negotiated wages w to wage-earners, they are making decision on how much of the remainder ($\nu k^{\alpha} - w$) they will consume and how much they will invest in capital growth.[2] We adopt a simple parametrisation where a constant fraction ρ_d is consumed by entrepreneurs, while the remainder ($1 - \rho_d$) is invested. Taking into account the

---

[1] It should be mentioned that the notation for w is slightly changed in the present book chapter as compared to previous publications on SDEM: if we denote by $\tilde{w}$ the 'old' wages from [12, 13], then w is related to $\tilde{w}$ as $w = (1 - \theta)\,\tilde{w}$, where θ is the fraction of entrepreneurs in the population of the region under study. With this change of variables, the parameter θ will no longer appear in SDEM model equations; see more details on θ in [12, 13].

[2] In [5, 12] the production function in SDEM was assumed to be linear, that is, of AK type (1), with $\alpha = 1$ in (6). In [13], the economy with increasing returns to scale was considered, with $\alpha > 1$ in (6).

capital depreciation and population growth exactly as was done in the AK model (2), we easily get the first dynamic equation (7a) of SDEM model.

The second dynamic equation (7b) describes the wage adjustment process (with the wage adjustment rate λ_W) driven by wage negotiations between wage-earners and entrepreneurs. Denote $w_{\text{targ}} = \nu k^\alpha - (\lambda_K + \lambda_L) k$; this term appears in the r.h.s. of (7b) and represents the target wage rate perceived by wage-earners as 'ideal'. If the actual wage rate were equal to w_{targ} that would mean that the maximal possible fraction of the output νk^α, except the investment $(\lambda_K + \lambda_L) k$ necessary to balance the depreciation of capital and population growth, would be channelled into wages. However, under such economic scenario both the growth rate of the economy and the consumption of entrepreneurs would be strictly zero. Obviously, such an equilibrium would not be optimal for entrepreneurs, and, as they have substantial negotiation power in wage negotiation process, wages in (7b) adjust not to w_{targ}, but rather to a lower wage rate $q w_{\text{targ}}$, where the parameter q $(0 < q < 1)$ describes the wage negotiation power of wage-earners (the stronger their power, the closer is q to unity, and hence the closer is $q w_{\text{targ}}$ to w_{targ}). Further details on the dynamical system (7a)–(7b) can be found in [5, 12, 13].

If the production function is linear ($\alpha = 1$), SDEM as described by a dynamical system (7a)–(7b), also becomes a linear model:

$$\dot{k} = [(1 - \rho_d) \nu - (\lambda_K + \lambda_L)] k - (1 - \rho_d) w, \qquad (8a)$$

$$\dot{w} = \lambda_W [q (\nu - (\lambda_K + \lambda_L)) k - w]. \qquad (8b)$$

We will show now that in the limit of immediate wage adjustment the SDEM model is formally reducible to the Solow-Swan model (or, in case of a linear production function, $\alpha = 1$, to the AK model).

Suppose that the wage adjustment is immediate, that is, $\lambda_W = +\infty$ in (7b). For consistency, the term in square brackets in (7b) should be then always strictly equal to zero; this yields

$$w \equiv q (\nu k^\alpha - (\lambda_K + \lambda_L) k). \qquad (9)$$

By substituting (9) in (7a) and rearranging the terms, we easily get

$$\dot{k} = (1 - \rho_d) (1 - q) \nu k^\alpha - (1 - q (1 - \rho_d)) (\lambda_K + \lambda_L) k. \qquad (10)$$

If we now introduce the notations

$$\tilde{s} \equiv 1 - \rho_d, \qquad (11a)$$

$$\tilde{A} \equiv (1 - q) \nu, \qquad (11b)$$

$$\tilde{\lambda}_K \equiv (1 - q (1 - \rho_d)) \lambda_K, \qquad (11c)$$

$$\tilde{\lambda}_L \equiv (1 - q (1 - \rho_d)) \lambda_L, \qquad (11d)$$

then we rewrite (10) to the form

$$\dot{k} = \tilde{s}\tilde{A}k^\alpha - (\tilde{\lambda}_K + \tilde{\lambda}_L)k, \tag{12}$$

formally equivalent to the standard Solow–Swan model (5), or to the AK model (2) in the limiting case $\alpha = 1$. Still, the analogy is somewhat formal, as, for instance, the 'objective' depreciation rate $\tilde{\lambda}_K$ in (11c) now depends on the policy parameter of the model ρ_d.

3 Analysis of Model Solutions

We now turn to the analysis of properties of solutions of four models introduced above. For the three members of our model family out of four (except the nonlinear SDEM model) it is possible to obtain exact analytical solutions in closed form. The discussion in the present section will serve as a basis for the analysis of more complex models with drought damage functions incorporated (see Sects. 4 and 5).

3.1 The AK Model

The AK model has been formalized in Sect. 2.1 as a first-order homogeneous linear ODE (2). Therefore, its solution is straightforward:

$$k = k_0 \exp(r_0 t), \tag{13}$$

where k_0 is the initial value of capital and

$$r_0 = sA - (\lambda_K + \lambda_L) \tag{14}$$

is the constant growth rate of the economy. It should be mentioned that for very small savings rates s (a case of underinvestment) r_0 will be negative, and the economy will be gradually declining. In a regular case ($r_0 > 0$) the AK model describes an unbounded growth.

3.2 The Solow–Swan Model

Unlike in the AK model with unbounded growth, in the Solow–Swan model the solution $k(t)$ converges at $t \to +\infty$ to the stable fixed point k_* that does not depend on the initial value of k. The asymptotic value k_* can be easily found by equating the r.h.s. of (5) to zero, that yields:

$$k_*^{1-\alpha} = \frac{sA}{\lambda_K + \lambda_L}. \tag{15}$$

For a particular case of Cobb–Douglas production function (4), the transitional dynamics of $k(t)$ in the Solow–Swan model (5) can be described analytically in closed form. Indeed, in this case (5) can be reduced to a linear ODE [23]. We will apply this method to a more complex case of the Solow–Swan model with drought damages in Sect. 4.2.2 below.

3.3 The SDEM Model

The linear SDEM model is described by a two-dimensional system of linear ODE (8a)–(8b), and therefore can be straightforwardly analysed by standard methods, while for the nonlinear SDEM the analytical treatment is possible only partially.

3.3.1 Linear SDEM Model

As discussed in Sect. 2.3, when $\alpha = 0$ SDEM becomes a linear dynamical model defined by (8a)–(8b). Given that the model is two-dimensional, it is possible to write down explicitly exact analytical solutions in linear case. However, we will limit ourselves to writing the solution of related matrix equation in symbolic form and then analysing its asymptotic properties.

Firstly, we introduce a vector of state variables

$$\mathbf{z} = \begin{pmatrix} k \\ w \end{pmatrix}. \tag{16}$$

Then, in accordance with (8a)–(8b), the matrix dynamic equation takes the form

$$\dot{\mathbf{z}} = \hat{\mathbf{A}}\mathbf{z}, \tag{17}$$

where a matrix

$$\hat{\mathbf{A}} = \begin{pmatrix} a_{11} & a_{12} \\ a_{21} & a_{22} \end{pmatrix} \tag{18}$$

has the following constant elements:

$$a_{11} = (1 - \rho_d)\, v - (\lambda_K + \lambda_L)\,, \tag{19a}$$
$$a_{12} = -(1 - \rho_d)\,, \tag{19b}$$
$$a_{21} = \lambda_W q\, (v - (\lambda_K + \lambda_L))\,, \tag{19c}$$
$$a_{22} = -\lambda_W. \tag{19d}$$

The solution of (17) in symbolic form is

$$\mathbf{z}(t) = \exp\left(\hat{\mathbf{A}}t\right)\mathbf{z}_0, \tag{20}$$

where $\mathbf{z}_0 = (k_0, w_0)^\top$ is the initial value of $\mathbf{z}$ at $t = 0$.

To analyse the asymptotic properties of the solution (20), we need to know the properties of eigenvalues of matrix $\hat{\mathbf{A}}$. The corresponding characteristic equation defining the eigenvalues is quadratic:

$$\det\left(\hat{\mathbf{A}} - \lambda\hat{\mathbf{I}}\right) = \lambda^2 - \text{Tr}\hat{\mathbf{A}} \cdot \lambda + \det\hat{\mathbf{A}} = 0, \tag{21}$$

with the coefficients

$$\text{Tr}\hat{\mathbf{A}} = \nu - (\lambda_K + \lambda_L + \lambda_W) - \rho_d \cdot \nu, \tag{22}$$

$$\det\hat{\mathbf{A}} = (1 - q)\lambda_W\left[\lambda_K + \lambda_L - \nu + \rho_d \cdot \left\{\nu + \frac{q}{1-q}(\lambda_K + \lambda_L)\right\}\right]. \tag{23}$$

Explicitly, the two roots of (21) are

$$\lambda_\pm = \frac{\text{Tr}\hat{\mathbf{A}}}{2} \pm \sqrt{\frac{\text{Tr}^2\hat{\mathbf{A}}}{4} - \det\hat{\mathbf{A}}}. \tag{24}$$

Consider (23) when ρ_d varies from zero to unity. Note that, normally, $\nu > \lambda_K + \lambda_L$, as, numerically, the technology parameter is usually order of magnitude higher than both the depreciation rate and the population growth rate. Then for $\rho_d = 0$, and, therefore, also for sufficiently small positive ρ_d, we have $\det\hat{\mathbf{A}} < 0$. Therefore, from (24) λ_+ and λ_- are real. Given that the roots are real, we immediately get from Vieta's formula $\lambda_+\lambda_- = \det\hat{\mathbf{A}}$ that they have opposite signs: the bigger root $\lambda_+ > 0$, while the smaller root $\lambda_- < 0$. With one positive root, at large times linear SDEM will be in a regime of economic growth (possibly, dependent on initial conditions, after a non-monotonous transitional period). This ensures a meaningful model setup. Therefore, in what follows, we will always assume that ρ_d is small enough to guarantee $\det\hat{\mathbf{A}} < 0$.

3.3.2 SDEM Model with Decreasing Returns to Scale

In case of nonlinear SDEM model (7a)–(7b), it is not possible to derive its exact analytical solution. However, paper-and-pencil analysis of the non-trivial fixed point of SDEM and of its stability can be performed.

It can be easily shown that the dynamical system (7a)–(7b) of SDEM model with decreasing returns to scale ($0 < \alpha < 1$) has two fixed points on the phase plane (k, w): (i) the trivial fixed point $(0, 0)$: $k = 0$, $w = 0$; (ii) the non-trivial fixed point (k_*, w_*) where

$$k_*^{1-\alpha} = \frac{\nu}{\lambda_K + \lambda_L} \frac{(1 - q)(1 - \rho_d)}{1 - q(1 - \rho_d)}, \tag{25a}$$

$$w_* = (\lambda_K + \lambda_L) \frac{q}{1 - q} \frac{\rho_d}{1 - \rho_d} k_*. \tag{25b}$$

Below we perform the Lyapunov stability analysis of the non-trivial fixed point, that in many aspects will resemble the stability analysis of SDEM with *increasing* returns to scale presented in [13], yet will lead to qualitatively different results.

The dynamics of linearized system in the neighbourhood of non-trivial fixed point (k_*, w_*) are described by an equation

$$\begin{pmatrix} \delta \dot{k} \\ \delta \dot{w} \end{pmatrix} = \hat{\mathbf{A}}_0 \begin{pmatrix} \delta k \\ \delta w \end{pmatrix}, \tag{26}$$

where the matrix

$$\hat{\mathbf{A}}_0 = \begin{pmatrix} a_{11}^0 & a_{12}^0 \\ a_{21}^0 & a_{22}^0 \end{pmatrix} \tag{27}$$

has the following elements:

$$a_{11}^0 = (1 - \rho_d) \nu \alpha k_*^{\alpha-1} - (\lambda_K + \lambda_L), \tag{28a}$$

$$a_{21}^0 = \lambda_W q \left[\nu \alpha k_*^{\alpha-1} - (\lambda_K + \lambda_L) \right], \tag{28b}$$

$$a_{12}^0 = -(1 - \rho_d), \tag{28c}$$

$$a_{22}^0 = -\lambda_W. \tag{28d}$$

Making use of (25a), the first column of the matrix $\hat{\mathbf{A}}_0$ ((28a), (28b)) can be rewritten as

$$a_{11}^0 = (\lambda_K + \lambda_L) \left[\frac{q}{1 - q} \alpha \rho_d - (1 - \alpha) \right], \tag{29a}$$

$$a_{21}^0 = (\lambda_K + \lambda_L) \lambda_W q \frac{\rho_d - (1 - \alpha)(1 - q(1 - \rho_d))}{(1 - q)(1 - \rho_d)}. \tag{29b}$$

The characteristic equation of the matrix $\hat{\mathbf{A}}_0$ has the form

$$\lambda^2 - \operatorname{Tr}\hat{\mathbf{A}}_0 \cdot \lambda + \det\hat{\mathbf{A}}_0 = 0 \tag{30}$$

with the coefficients

$$\text{Tr}\hat{\mathbf{A}}_0 = (\lambda_K + \lambda_L)\left[\frac{q}{1-q}\alpha\rho_d - (1-\alpha)\right] - \lambda_W, \tag{31}$$

$$\det\hat{\mathbf{A}}_0 = (\lambda_K + \lambda_L)\,\lambda_W\,(1-\alpha)\,(1-q\,(1-\rho_d))\,. \tag{32}$$

To ensure the stability of the fixed point (25a)–(25b), the characteristic polynomial in (30) should be Hurwitz polynomial, i.e. a polynomial whose roots have negative real parts. It is known that a quadratic polynomial of the kind $\lambda^2 + p\lambda + q$ has both its roots with negative real parts if and only if both p and q are positive [24]. In particular, for the characteristic polynomial in (30) this criterion takes the form

$$\text{Tr}\hat{\mathbf{A}}_0 < 0, \qquad \det\hat{\mathbf{A}}_0 > 0. \tag{33}$$

In accordance with (32), the second inequality ($\det\hat{\mathbf{A}}_0 > 0$) always holds. Whether the first inequality ($\text{Tr}\hat{\mathbf{A}}_0 < 0$) is valid depends on the values of model parameters. However, in any case, at least for sufficiently small values of ρ_d,

$$0 < \rho_d < \rho_d^*, \tag{34}$$

where

$$\rho_d^* = \frac{1-q}{\alpha q}\left[\frac{\lambda_W}{\lambda_K + \lambda_L} + 1 - \alpha\right], \tag{35}$$

the first inequality in (33) also holds. It should be stressed that, dependent on the values of model parameters, ρ_d^* in (35) might be greater than unity; in such a case, the condition $\text{Tr}\hat{\mathbf{A}}_0 < 0$ holds for any admissible values of ρ_d ($0 < \rho_d < 1$).

To summarize, the characteristic polynomial in (30) will be Hurwitz polynomial, and hence the non-trivial fixed point of nonlinear SDEM model will be stable, at least for sufficiently small ρ_d, (34)–(35); dependent on model parameters, the non-trivial fixed point can be stable for any admissible values of ρ_d ($0 < \rho_d < 1$). In our further analysis of SDEM model, we will always assume the stability of the non-trivial fixed point. This also means, that in the limit $t \to +\infty$ model solutions will converge to this fixed point.

4 Economic Dynamics with Deterministic Drought Damages

Based on the preceding discussion in Sects. 2–3, we will now incorporate economic impacts of droughts in the modelling scheme.

Common to many models of climate and environmental economics [8, 22, 33, 34], we will assume that climate-related damages (in the context of the present study, drought damages) from which the regional economy suffers might be parametrised by *drought damage function* $d(t)$. This concept is common to a number of studies on drought impact modelling and drought risk assessment [3, 10, 19]. Below we assume that the drought damage function $d(t)$ is exogenous and, in general, time-dependent. Drought damages affect the modelled economy in the following way: it is assumed that a fraction $d(t)$ $(0 \leq d(t) \leq 1)$ of regional output is lost at time moment t as drought damages, and only the remainder,

$$y_{\text{eff}}(t) = [1 - d(t)]\, y(t), \qquad 0 \leq d(t) \leq 1, \tag{36}$$

is available for consumption or investment.

Drought damage functions might be specified either as deterministic functions or as random processes. In the former case, they are useful to simulate the 'smooth' dynamics of water-constrained economies over long time horizons. In the latter case, short-term variability of drought conditions is taken into account. In the present section, we will focus on deterministic drought damage functions, postponing the consideration of stochastic $d(t)$ until Sect. 5.

Specifically, we will assume a model

$$d(t) = d_0(1 + \Delta t), \qquad 0 \leq d_0 \leq 1, \qquad \Delta \geq 0, \tag{37}$$

where $d_0 = \text{const}$ is the initial value of drought damages and $\Delta = \text{const}$ is damage growth rate. In a particular case of $\Delta = 0$ the damages are constant (Sect. 4.1 below); otherwise ($\Delta > 0$) damages are growing over time linearly (Sect. 4.2 below). In case of linear growth, the model (37) is applicable only within a finite time horizon limited by an obvious constraint $d(t) \leq 1$ (damage losses cannot exceed 100 per cent of output). Specifically, the model (37) can be applied at $0 \leq t \leq t_{\text{end}}$ where

$$t_{\text{end}} = \frac{1}{\Delta}\left[\frac{1}{d_0} - 1\right]. \tag{38}$$

The analysis of regional economic dynamics with deterministic drought damages gradually growing over time would provide first insights to projected future impacts of climate change on a water-constrained economy, as, according to climate model projections, frequency and severity of droughts is expected to increase under changing climate in many regions of the world [9, 18, 30].

4.1 Constant Deterministic Damages

In the case of constant drought damages, $d(t) = d_0 = \text{const}$, $\Delta = 0$, all models introduced in Sect. 2 remain applicable if we rescale (decrease) the technology parameter relating physical capital to output: A in case of the AK and the Solow–Swan models, v in case of linear and nonlinear SDEM models. According to (36), the conver-

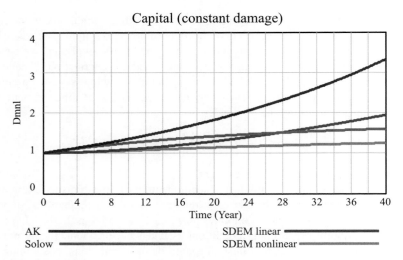

Fig. 1 Capital dynamics in case of constant drought damages: the AK model; the Solow–Swan model; linear SDEM model; nonlinear SDEM model

sion factor is equal to $(1 - d_0)$, so we should replace A with $(1 - d_0)A$ and ν with $(1 - d_0)\nu$. After such rescaling, all results derived in Sect. 3 above (in particular, analytical solutions and fixed points) remain valid. Obviously, drought damages slow down economic growth and (in case of nonlinear models under consideration) lead to equilibria with lower output and consumption.

Figures 1 and 2 visualise the dynamics of water-constrained economy under constant deterministic damages calculated with our model family. Capital dynamics (that are calculated by all four models) are shown in Fig. 1, while wage dynamics (calculated by linear and nonlinear SDEM only) are shown in Fig. 2.

The following numerical values of common model parameters and initial conditions are chosen: $\lambda_K = 0.05$ year^{-1}, $\lambda_L = 0.01$ year^{-1}, $\nu = A = 0.5$ year^{-1}, $d_0 = 0.1$, $k_0 = 1.0$. For the Solow-Swan model and the nonlinear SDEM model, $\alpha = 0.25$. For linear and nonlinear SDEM models, $\lambda_W = 0.2$ year^{-1}, $\rho_d = 0.6$, $q = 0.7$, $w_0 = 0.3$ year^{-1}.

As is seen on Figs. 1 and 2, for the values of model parameters chosen, state variables experience unbounded growth for linear models (the AK model, the linear SDEM model), while for nonlinear models (the Solow–Swan model, the nonlinear SDEM model) they converge to their asymptotic values predicted by the theory developed in Sect. 3.

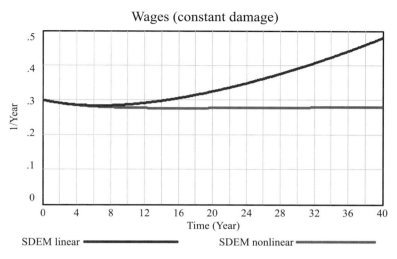

Fig. 2 Wage dynamics in case of constant drought damages: linear SDEM model; nonlinear SDEM model

4.2 Deterministic Damages Growing over Time

We now turn to the case of linear growth of drought damages (37), $\Delta > 0$. Although economic dynamics become more complex, it is still possible to derive exact analytical solutions for the AK model and for the Solow–Swan model with Cobb–Douglas production function. Unfortunately, both linear and nonlinear SDEM models are no longer analytically tractable under time-dependent drought damages.

4.2.1 The AK Model

As discussed above, the AK model with time-dependent drought damages takes the form

$$\dot{k} = [(1 - d(t)) sA - (\lambda_K + \lambda_L)] k. \tag{39}$$

The dynamic equation (39) is obtained from (2) by replacing A with $(1 - d(t)) A$. This equation can be easily integrated, that yields

$$k(t) = k_0 \exp\left[r_0 t - sA \int_0^t d\left(t'\right) dt' \right], \tag{40}$$

where r_0 is the same baseline growth rate as in Eq. (13).

In a particular case of linear damage growth (37), the solution takes the form

$$k(t) = k_0 \exp\left[(r_0 - d_0 s A) \cdot t - \frac{1}{2} d_0 \Delta s A \cdot t^2 \right]. \tag{41}$$

As discussed above, the solution (41) is valid only for $0 \leq t \leq t_{\text{end}}$ where t_{end} is defined by (37).

As is seen from (41), the unbounded growth of the standard AK model is no longer the case with growing climate damages. Instead, in long-term the economy goes to the regime of decay.

4.2.2   The Solow–Swan Model

Quite analogously to the consideration from the previous section, the Solow–Swan model with time-dependent drought damages takes the form

$$\dot{k} = (1 - d(t)) s A k^\alpha - (\lambda_K + \lambda_L) k. \tag{42}$$

As mentioned in Sect. 3.2, in a particular case of Cobb–Douglas production function (4) the standard Solow–Swan model (5) can be solved analytically. Specifically, by change of variable, it can be reduced to a linear ordinary differential equation [23]. We apply the same recipe to the damage case (42) and define an auxiliary variable

$$z = k^{1-\alpha}. \tag{43}$$

By substituting (43) into the nonlinear equation (42) and making some rearrangements, we easily get a linear equation

$$\dot{z} + (1 - \alpha)(\lambda_K + \lambda_L) z = (1 - \alpha) s A (1 - d(t)). \tag{44}$$

Denote for brevity an auxiliary parameter

$$b = (1 - \alpha)(\lambda_K + \lambda_L), \tag{45}$$

then Eq. (44) can be rewritten in the form

$$\dot{z} + bz = (1 - \alpha) s A (1 - d(t)). \tag{46}$$

With known time dependence of drought damages $d(t)$, the transformed equation (46) can be integrated, and its solution takes the form

$$z(t) = \exp(-bt)\left[z_0 + (1 - \alpha) s A \int_0^t (1 - d(t')) \exp(bt') \, dt' \right], \tag{47}$$

where z_0 is the initial value of z:

$$z_0 = k_0^{1-\alpha}. \tag{48}$$

After the integral is taken in Eq. (47), $k(t)$ can then be found from Eq. (43).

In a particular case of linearly growing drought damages (37), the solution (47) takes the form

$$z(t) = [g_0 - d_0 \Delta bt] \frac{(1 - \alpha) sA}{b^2} + \left[z_0 - g_0 \frac{(1 - \alpha) sA}{b^2} \right] \exp(-bt), \tag{49}$$

where another auxiliary constant

$$g_0 = (1 - d_0) b + d_0 \Delta \tag{50}$$

is introduced for brevity.

Figures 3 and 4 show capital and wage dynamics simulated by our model family for deterministic drought damages growing over time. The numerical value $\Delta = 0.1$ year^{-1} is chosen for the growth rate in drought damage function (37). The numerical values of all other parameters are the same as specified at the end of Sect. 4.1 for Figs. 1 and 2. According to (38), the end date of applicability of the damage function (37) $t_{\text{end}} = 90$ years in this case; this is far later than the end date of simulations presented in Figs. 3 and 4. As opposed to Figs. 1 and 2, Figs. 3 and 4 show the transition to decaying regime in the long term for all models.

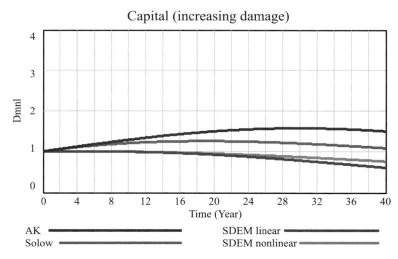

Fig. 3 Capital dynamics in case of deterministic drought damages growing over time: the AK model; the Solow–Swan model; linear SDEM model; nonlinear SDEM model

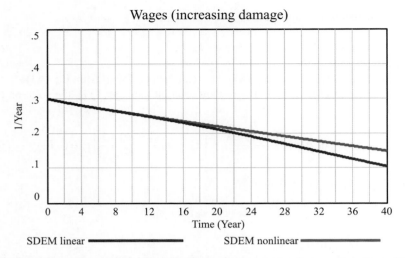

Fig. 4 Wage dynamics in case of deterministic drought damages growing over time: linear SDEM model; nonlinear SDEM model

5   Introducing the Stochasticity in Drought Damages

So far, our consideration of regional economic dynamics was purely deterministic. In particular, the drought damage function specified by (37) was deterministic. As discussed in Sect. 4, this might be a reasonable approximation for modelling the 'smooth' long-term dynamics of water-constrained economies. At the same time, droughts are natural hazards, and regional drought occurrence and drought character-istics experience significant inter-annual variability. Therefore, for a deeper analysis of economic dynamics affected by droughts, it is important to include the stochastic component in drought damage function.

In the present section, we perform the 'stochastization' of two linear models considered above: the AK model and the linear SDEM model. Some comments on stochastic versions of nonlinear models will be provided at the end of the section.

Our analysis will be based on the theory of linear stochastic systems with paramet-ric noise, that allows deriving exact dynamical equations for the mean and covariance of the state vector in closed form. The general form of a linear stochastic system with parametric noise is [25]

$$\dot{\mathbf{z}} = \hat{\mathbf{A}}\mathbf{z} + \hat{\mathbf{a}}_0 + \left(\hat{\mathbf{B}}_0 + \sum_{h=1}^{p} \hat{\mathbf{B}}_h z_h, \right) \mathbf{v}, \tag{51}$$

where $\mathbf{z}$ is a p-dimensional state vector, z_h is its h-th component, $\hat{\mathbf{A}}$, $\hat{\mathbf{B}}_0$, $\hat{\mathbf{B}}_h$—given matrices (in general case, time-dependent), $\hat{\mathbf{a}}_0$—a given vector (in general case, also time-dependent), $\mathbf{v}$—the vector white noise of (generally, time-dependent) spectral density $\hat{v}$.

As is shown in [25], the differential equations for the mean $\mathbf{m}(t) = \langle \mathbf{z}(t) \rangle$ and covariance $\hat{\mathbf{K}}(t) = \langle (\mathbf{z}(t) - \mathbf{m}(t)) \left(\mathbf{z}^\top(t) - \mathbf{m}^\top(t) \right) \rangle$ take the closed form for the system (51):

$$\dot{\mathbf{m}} = \hat{\mathbf{A}}\mathbf{m} + \hat{\mathbf{a}}_0, \tag{52}$$

$$\dot{\hat{\mathbf{K}}} = \hat{\mathbf{A}}\hat{\mathbf{K}} + \hat{\mathbf{K}}\hat{\mathbf{A}}^\top + \hat{\mathbf{B}}_0\hat{v}\hat{\mathbf{B}}_0^\top + \sum_{h=1}^{p} \left(\hat{\mathbf{B}}_h\hat{v}\hat{\mathbf{B}}_0^\top + \hat{\mathbf{B}}_0\hat{v}\hat{\mathbf{B}}_h^\top \right) m_h + \sum_{h=1}^{p} \hat{\mathbf{B}}_h\hat{v}\hat{\mathbf{B}}_l^\top \left(m_h m_l + K_{hl} \right). \tag{53}$$

In (53), m_h and K_{hl} are the elements of the vector $\mathbf{m}$ and the matrix $\hat{\mathbf{K}}$, respectively.

According to (52), the dynamics of the mean are merely described by dynamics of the system unperturbed by the noise. The dynamics of covariance follow more complicated pattern.

Armed with this general result, we will now consider the AK model and the linear SDEM model with drought damages that include stochastic component. As deterministic part of the damage function, we will take (37) in a particular case of $\Delta = 0$. That means, we will model drought damages as a stationary random process with constant mean d_0 on which a random noise of constant spectral density is superimposed:

$$d(t) = d_0 \left(1 + \eta V \right), \tag{54}$$

where $\eta = $ const and, from now on, V is a scalar white noise of unit spectral density.

It should be noted that, from a perspective of hydrology, the white noise approximation for describing the economic impacts of droughts might be of limited applicability, as the stochastic dynamics of droughts exert more sophisticated patterns [18, 30]. Nevertheless, we believe that the models developed below might be a reasonable starting point in describing stochastic drought damages.

5.1 Stochastic AK Model

Analogously to the deterministic case (39), the AK model with stochastic drought damages (54) might be written as

$$\dot{k} = [(1 - d_0) \, sA - (\lambda_K + \lambda_L)] k - d_0 s A \eta V k. \tag{55}$$

Denote, for brevity, auxiliary constants

$$a = (1 - d_0) \, sA - (\lambda_K + \lambda_L), \tag{56a}$$
$$b = -d_0 s A \eta. \tag{56b}$$

Then (55) can be concisely rewritten as

$$\dot{k} = ak + bkV. \tag{57}$$

Note that (57) is a particular case of scalar equation (51) with $a_0 = 0$, $b_0 = 0$. Therefore, we can immediately write related scalar equations for the mean $m(t)$ and variance $K(t)$:

$$\dot{m} = am, \tag{58}$$

$$\dot{K} = 2aK + b^2 \left(m^2 + K\right). \tag{59}$$

Assuming that the initial condition for k is certain ($m_0 = k_0$ at $t = 0$), we firstly solve (58) easily,

$$m(t) = k_0 \exp(at). \tag{60}$$

Then, by substituting the solution (60) into (59), we get an equation for K

$$\dot{K} = 2aK + b^2 \left(k_0^2 \exp(2at) + K\right). \tag{61}$$

Its particular solution corresponding to certainty at initial time moment ($K_0 = 0$ at $t = 0$) takes the form

$$K(t) = m_0^2 \left[\exp\left(\left(2a + b^2\right)t\right) - \exp(2at)\right]. \tag{62}$$

The standard deviation of capital, $\sigma_K(t) = \sqrt{K(t)}$, is then equal to

$$\sigma_K(t) = m_0 \sqrt{\exp\left(\left(2a + b^2\right)t\right) - \exp(2at)}. \tag{63}$$

Figure 5 presents capital dynamics simulated by the AK model with the stochastic component of drought damage function. The dynamics of the mean value $m(t)$ (60), confidence intervals ($m(t) - \sigma_K(t)$; $m(t) + \sigma_K(t)$), where $\sigma_K(t)$ is given by (63), and three realizations of random capital dynamics are shown. Random realizations are computed by the Euler-Maruyama method [27]. The numerical value $\eta = 2.0$ is chosen for simulations; the numerical values of all other parameters are the same as specified at the end of Sect. 4.1. Overall, the stochastic dynamics clearly manifest the growing uncertainty of economic projections caused by stochasticity of drought damages.

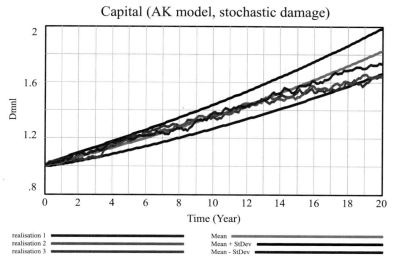

Fig. 5 Capital dynamics simulated by the AK model with the stochastic component of drought damage function. The dynamics of the mean value, confidence intervals and three realizations of random capital dynamics are shown

5.2 Stochastic Linear SDEM Model

Using the notation (16) for the state vector, we might write the linear stochastic SDEM model in the form[3]

$$\dot{\mathbf{z}} = \hat{\mathbf{A}}_d \mathbf{z} + \hat{\mathbf{B}}_1 z_1 \mathbf{v}, \tag{64}$$

where a matrix

$$\hat{\mathbf{A}}_d = \begin{pmatrix} a_{11}^d & a_{12}^d \\ a_{21}^d & a_{22}^d \end{pmatrix} \tag{65}$$

has the following constant elements:

$$a_{11}^d = (1 - \rho_d)\,(1 - d_0)\,v - (\lambda_K + \lambda_L)\,, \tag{66a}$$
$$a_{12}^d = -\,(1 - \rho_d)\,, \tag{66b}$$
$$a_{21}^d = \lambda_W q\,[(1 - d_0)\,v - (\lambda_K + \lambda_L)]\,, \tag{66c}$$
$$a_{22}^d = -\lambda_W\,, \tag{66d}$$

$z_1 \equiv k$, and a new auxiliary constant matrix $\hat{\mathbf{B}}_1$ is introduced:

---

[3] A different stochastic version of linear SDEM model was considered in [12]. There a parametric noise was caused by fluctuations of the wage negotiation parameter q.

$$\hat{\mathbf{B}}_1 = -v d_0 \eta \begin{pmatrix} 1 - \rho_d & 0 \\ \lambda_w q & 0 \end{pmatrix}. \tag{67}$$

The vector white noise in (64) has the first white noise component with unit spectral density, while its second component is equal to zero:

$$\mathbf{v} = \begin{pmatrix} V \\ 0 \end{pmatrix}. \tag{68}$$

The stochastic system in (64) is a particular case of (51) with $\hat{\mathbf{A}} = \hat{\mathbf{A}}_d$, $\hat{\mathbf{a}}_0 = \mathbf{0}$, $\hat{\mathbf{B}}_0 = \mathbf{0}$, and the related equations for the mean and variance take the form

$$\dot{\mathbf{m}} = \hat{\mathbf{A}}_d \mathbf{m}, \tag{69}$$

$$\dot{\hat{\mathbf{K}}} = \hat{\mathbf{A}}_d \hat{\mathbf{K}} + \hat{\mathbf{K}} \hat{\mathbf{A}}_d^\top + \hat{\mathbf{B}}_1 \hat{\mathbf{B}}_1^\top \left(m_1^2 + K_{11} \right). \tag{70}$$

The solution of (69) in symbolic form is

$$\mathbf{m}(t) = \exp \left(\hat{\mathbf{A}}_d t \right) \mathbf{m}_0. \tag{71}$$

By introducing another auxiliary constant matrix

$$\hat{\mathbf{D}} \equiv \hat{\mathbf{B}}_1 \hat{\mathbf{B}}_l^\top = v^2 d_0^2 \eta^2 \begin{pmatrix} (1 - \rho_d)^2 & (1 - \rho_d) \lambda_w q \\ (1 - \rho_d) \lambda_w q & \lambda_w^2 q^2 \end{pmatrix}, \tag{72}$$

we can explicitly write the scalar equations for individual elements of matrix $\hat{\mathbf{K}}$ in the form.

$$\dot{K}_{11} = \left(2a_{11}^d + D_{11} \right) K_{11} + 2a_{12}^d K_{12} + D_{11} m_1^2, \tag{73a}$$

$$\dot{K}_{12} = \left(a_{21}^d + D_{12} \right) K_{11} + \left(a_{11}^d + a_{22} \right) K_{12} + a_{12}^d K_{22} + D_{12} m_1^2, \tag{73b}$$

$$\dot{K}_{22} = D_{22} K_{11} + 2a_{21}^d K_{12} + 2a_{22}^d K_{22} + D_{22} m_1^2. \tag{73c}$$

The fourth equation for the dynamics of K_{21} is unnecessary, because of the variance symmetry, $K_{21} = K_{12}$ (this property has been also used when writing down dynamical equations (73a)–(73c)).

When the system (73a)–(73c) is solved, the dynamics of standard deviations for capital ($\sigma_K(t)$) and wages ($\sigma_W(t)$) are given by identities

$$\sigma_k(t) = \sqrt{K_{11}(t)}, \qquad \sigma_W(t) = \sqrt{K_{22}(t)}. \tag{74}$$

Capital and wage dynamics simulated by the linear SDEM model with the stochastic component of drought damage function are presented on Figs. 6 and 7, respectively. Comments analogous to those made at the end of Sect. 5.1 apply in this case as well.

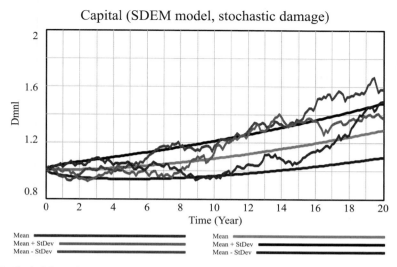

Fig. 6 Capital dynamics simulated by linear SDEM model with the stochastic component of drought damage function. The dynamics of the mean value, confidence intervals and three realizations of random capital dynamics are shown

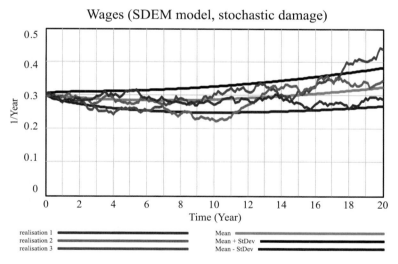

Fig. 7 Wage dynamics simulated by linear SDEM model with the stochastic component of drought damage function. The dynamics of the mean value, confidence intervals and three realizations of random wage dynamics are shown

Unfortunately, in case of nonlinear stochastic systems we cannot proceed with analytical treatment as far as in the linear case. Consider, for instance, the Solow–Swan model with stochastic drought damages. There might be a temptation to consider the stochastic analogue of linear ODE (44), to which, as discussed in Sect. 4.2.2, the nonlinear Solow–Swan model with deterministic damages (42) is reducible by changing of variable. Methods described in the present section would be of course applicable to linear stochastic analogue of (44). However, a closer examination reveals that this linear stochastic ODE would *not* be equivalent to the initial nonlinear stochastic Solow–Swan model. Indeed, a change of variable analogous to (43) would lead, in accordance with the Itô formula [25, 27], to a *nonlinear* stochastic differential equation. However, it turns out that for the stochastic Solow–Swan model with the Cobb–Douglas production function it is possible to calculate the steady-state probability density function by methods analogous to those applied by [17].

6 Conclusions

Modelling of water-constrained economies affected by droughts performed in the present study leads to several general conclusions.

Firstly, we intentionally performed simulations with a family of models, not with a single model. Members of model family resemble each other in certain aspects, share common numerical values of many parameters, and they describe economic dynamics broadly at the same level of complexity. However, simulations with these models yield qualitatively and quantitatively different solutions. This is especially true for long-term asymptotic behaviour of basic models (Sects. 2–3), before drought damage functions have been included in the modelling scheme. Therefore, to achieve reliable results, special care should be taken when selecting particular structure of the model and estimating its parameters versus real-world data.

Secondly, models with drought damages incorporated are sensitive to assumptions on the structure and values of parameters of damage functions. Scenarios with constant damages (Sect. 4.1, Figs. 1 and 2) differ fundamentally from scenarios with damages growing over time (Sect. 4.2, Figs. 3 and 4). As the latter reflect more precisely the projected increase of frequency and severity of droughts under climate change, in the domain of deterministic modelling preference should be given to models with steadily increasing damages.

Thirdly, including stochasticity in models of water-constrained economies is important. Although linear stochastic models with parametric noise developed in Sect. 5 describe the same mean dynamics as their deterministic counterparts, their added values is in the detailed description of uncertainty caused by drought variability (Figs. 5, 6 and 7). The nonlinear stochastic models might simulate the 'mean' dynamics in a substantially different way than their 'averaged' deterministic analogues, and further research in this direction opens interesting prospects. As mentioned above, it is also important to go beyond white noise approximation in stochastic drought damages, and to replace them by random processes with more realistic statistical properties.

Finally, and most importantly for policy implications, all models considered in the present study are 'business-as-usual' models, in the conventional terminology of economics of climate change [22, 33]: their policy variables (the savings rate s in the AK and the Solow–Swan models, the entrepreneur dividend parameter ρ_d in linear and nonlinear SDEM models) are kept constant (inflexible) during the simulation runs, and no options for adaptation to droughts are included in the modelling scheme. Modelling the adaptation options would alter the dynamics of water-constrained economies, and extended models allowing for adaptation options definitely deserve even more detailed analysis than presented by us in this study.

Acknowledgements The research leading to the results presented in the book chapter was supported by EU H2020 IMPREX project (Grant Agreement No. 641811). The first results of numerical simulations with stochastic SDEM model were presented earlier in IMPREX project report [16] and in a conference paper [14]. The authors are indebted to many participants of IMPREX project and of the 36th International Conference of the System Dynamics Society (Reykjavík, Iceland, August 6–10, 2018) for helpful comments.

References

1. Afraimovich VS, Rabinovich MI, Varona P (2004) Heteroclinic contours in neural ensembles and the winnerless competition principle. Int J Bifurc Chaos 14(04):1195–1208
2. Afraimovich VS, Zhigulin VP, Rabinovich MI (2004) On the origin of reproducible sequential activity in neural circuits. Chaos Interdiscip J Nonlinear Sci 14(4):1123–1129
3. Bachmair S, Svensson C, Prosdocimi I, Hannaford J, Stahl K (2017) Developing drought impact functions for drought risk management. Nat Hazards Earth Syst Sci 17:1947–1960
4. Barro RJ, Sala-i Martin XI (2003) Economic Growth, 2nd edn. The MIT Press
5. Barth V (2003) Integrated assessment of climate change using structural dynamic models. PhD Thesis, Max-Planck-Institut für Meteorologie, Hamburg
6. Barth V, Hasselmann K (2005) Analysis of climate damage abatement costs using a dynamic economic model. Vierteljahresheft zur Wirtschaftsforschung (DIW) 74(2):148–163
7. Carmona M, Máñez Costa M, Andreu J, Pulido-Velazquez M, Haro-Monteagudo D, Lopez-Nicolas A, Cremades R (2017) Assessing the effectiveness of multi-sector partnerships to manage droughts: the case of the Jucar river basin. Earth's Future 5(7):750–770
8. Hallegatte S, Ghil M (2008) Natural disasters impacting a macroeconomic model with endogenous dynamics. Ecol Econ 68(1):582–592
9. van den Hurk BJ, Bouwer LM, Buontempo C, Döscher R, Ercin E, Hananel C, Hunink JE, Kjellström E, Klein B, Manez M, Pappenberger F, Pouget L, Ramos MH, Ward PJ, Weerts AH, Wijngaard JB (2016) Improving predictions and management of hydrological extremes through climate services
10. Jenkins K, Warren R (2015) Drought-damage functions for the estimation of drought costs under future projections of climate change. J Extrem Events 02(01):1550,001

11. Knight FH (1944) Diminishing returns from investment. J Polit Econ 52:26–47
12. Kovalevsky DV (2014) Balanced growth in the Structural Dynamic Economic Model SDEM-2. Discontinuity Nonlinearity Complex 3(3):237–253
13. Kovalevsky DV (2016) Introducing increasing returns to scale and endogenous technological progress in the Structural Dynamic Economic Model SDEM-2. Discontinuity Nonlinearity Complex 5(1):1–8
14. Máñez-Costa M, Kovalevsky DV (2018) Participatory system dynamics modelling for adaptation to extreme hydrological events under conditions of climate change. In: Record of the 36th International Conference of the System Dynamics Society, Reykjavík, Iceland, August 7–9, 2018
15. Máñez-Costa M, Carmona M, Gerkensmeier B (2014) Assessing Governance Performance, Report 20. Climate Service Center, Germany
16. Máñez-Costa M, Osorio J, Kovalevsky D (2017) Generic integrative modeling approach guideline. EU H2020 IMPREX project deliverable D13.1
17. Merton RC (1975) An asymptotic theory of growth under uncertainty. Rev Econ Stud 42(3):375–393
18. Mishra AK, Singh VP (2010) A review of drought concepts. J Hydrol 391(1):202–216
19. Naumann G, Spinoni J, Vogt JV, Barbosa P (2015) Assessment of drought damages and their uncertainties in Europe. Environ Res Lett 10(12):124,013
20. von Neumann J (1937) Über ein Ökonomisches Gleichungssystem und eine Verallgemeinerung des Brouwerschen. Ergebnisse eines Mathematische Kolloquiums 8
21. von Neumann J (1945) A model of general equilibrium. Rev Econ Stud 13:1–9
22. Nordhaus WD (2008) A Question of Balance. Yale University Press, New Haven & London
23. Novales A, Fernández E, Ruíz J (2014) Economic Growth: Theory and Numerical Solution Methods, 2nd edn. Springer-Verlag, Berlin, Heidelberg
24. Petrov YP (1987) Synthesis of Optimal Control Systems under Incompletely Known Perturbing Forces [Sintez Optimal'nykh Sistem Upravlenija pri Nepolnost'ju Izvestnykh Vozmushhajushhikh Silakh] (in Russian). LGU Publishing House
25. Pugachev VS, Sinitsyn IN (1987) Stochastic Differential Systems Analysis and Filtering. Wiley, Chichester
26. Rabinovich MI, Huerta R, Varona P, Afraimovich VS (2008) Transient cognitive dynamics, metastability, and decision making. PLoS Comput Biol 4(5):e1000,072
27. Särkkä S (2012) Applied Stochastic Differential Equations. Written material for the course held in Autumn 2012
28. Solow RM (1956) A contribution to the theory of economic growth. Quart J Econ 70:65–94
29. Swan TW (1956) Economic growth and capital accumulation. Econ Record 32:334–361
30. Vicente-Serrano SM, Beguería S, López-Moreno JI (2010) A multiscalar drought index sensitive to global warming: the standardized precipitation evapotranspiration index. J Clim 23(7):1696–1718
31. Volchenkov D (2016) Survival Under Uncertainty. An Introduction to Probability Models of Social Structure and Evolution. Springer International Publishing
32. Volchenkov D, Helbach J, Tscherepanow M, Küheel S (2013) Exploration-exploitation trade-off in a treasure hunting game. Electron Notes Theor Comput Sci 299:101–121
33. Weber M, Barth V, Hasselmann K (2005) A multi-actor dynamic integrated assessment model (MADIAM) of induced technological change and sustainable economic growth. Ecol Econ 54(2):306–327

34. Weitzman ML (2012) GHG targets as insurance against catastrophic climate damages. J Publ Econ Theory 14(2):221–244
35. Williams DS, Máñez Costa M, Celliers L, Sutherland C (2018) Informal settlements and flooding: identifying strengths and weaknesses in local governance for water management. Water 10:871

An Efficient Operational Matrix Technique for Variable-Order Fractional Optimal Control Problems

H. Hassani, J. A. Tenreiro Machado, and Z. Avazzadeh

Abstract This chapter considers a class of variable-order fractional optimal control problems (V-FOCP). An optimization method based on the generalized polynomials (GP) for solving V-FOCP is proposed. The solution of the problem is expanded in terms of the GP with some free coefficients (FC) and control parameters (CP). The FC and CP are obtained optimally by minimizing the error of the approximate solution. Furthermore, a new variable-order fractional operational matrix (V-FOM) in the Caputo sense for the GP is derived. The efficiency and accuracy of the method are demonstrated with some examples.

Keywords Variable-order fractional optimal control problems · Variable-order fractional operational matrix · Generalized polynomials · Free coefficients · Control parameters

1 Introduction

Fractional differential equations (FDE) generalize the classical differential equations with integer orders. The existence and uniqueness of the FDE solutions were investigated in the seminal works by Samko et al., Podlubny and Kilbas et al. [1–3]. Applications of FDE address a variety of disciplines. The FDE are used in biophysics, bioengineering, quantum mechanics, finance, control theory, image and sig-

H. Hassani (✉)
Faculty of Mathematics and Statistics, Ton Duc Thang University, Ho Chi Minh City, Vietnam
e-mail: hosseinhassani@tdtu.edu.vn

J. A. Tenreiro Machado
Department of Electrical Engineering, Institute of Engineering, Polytechnic of Porto, Porto, Portugal
e-mail: jtm@isep.ipp.pt

R. Dr. António Bernardino de Almeida, 4249-015 Porto, Portugal

Z. Avazzadeh
Department of Applied Mathematics, Xi'an Jiaotong-Liverpool University, Suzhou 215123, Jiangsu, China

131

nal processing [4–6]. Most FDE do not have exact analytic solutions and, therefore, approximate and numerical techniques must be used. Several numerical approaches have been proposed to obtain the approximate solutions of FDE such as, the finite difference [7, 8], homotopy perturbation [9, 10], wavelets [11–15], Adomian decomposition [16, 17], multistep [18, 19], predictor-corrector [20, 21], and tau [22, 23] methods.

Variable-order (VO) fractional derivatives generalize the fixed-order fractional derivatives. In fact, VO derivatives are operators whose order may vary as a continuous function of time or space. Some authors introduced different definitions of VO differential operators, each with a specific goal in mind. We have definitions such as the Riemann-Liouville, Grünwald-Letnikov, Caputo, Riesz and the Coimbra formulations [24–27]. We find problems in physics, biology, engineering, and finance that can be described efficiently by models using VO fractional calculus, such as in mechanical systems [28], diffusion processes [29], multi fractional Gaussian noise [30], and FIR filters [31]. However, the analytical solutions of VO fractional differential equations (V-FDE) are more difficult to obtain and, therefore, numerical techniques must be used. Dahaghin and Hassani [32] proposed an optimization operational matrix (OM) method based on the generalized polynomials (GP) for the nonlinear VO time fractional diffusion-wave equation. Sun et al. [33, 34] presented numerical methods based on finite difference schemes. Chen et al. [35, 36] discussed a class of numerical methods based on operational matrices of Bernstein polynomials for the VO fractional linear cable and fractional diffusion equations. Chen et al. [37] solved a class of nonlinear V-FDE using the Legendre wavelets. Zayernouri and Karniadakis [38] developed fractional spectral collocation methods for the linear and nonlinear VO fractional partial differential equations. Zhao et al. [39] presented second-order approximations for VO fractional derivatives with algorithms and applications. Yaghoobi et al. [40] proposed an efficient cubic spline approximation for V-FDE with time delay. Parsa et al. [41] employed extended algorithms for approximating VO fractional derivatives and applications. Parsa et al. [42] applied also an integro quadratic spline approach for a class of VO fractional initial value problems. Heydari et al. [43] proposed an optimization wavelet method for multi V-FDE. Tayebi et al. [44] developed meshless method based on the moving least squares approximation and the finite difference scheme for solving two-dimensional VO time fractional advection-diffusion equation. Several other studies on V-FDE can be found in [45–48].

Optimal control theory is a branch of optimization theory focused on minimizing a cost or maximizing a payoff. The fractional optimal control theory is a relatively new area in mathematics and engineering disciplines [49, 50]. Fractional optimal control problems (FOCP) can be expressed using different definitions of fractional derivatives, such as by means of the Riemann-Liouville and Caputo fractional formulations. It is well known that the analytical solution of the FOCP generally does not exist except for special cases. Therefore, numerical methods are often preferred to obtain an approximate solution for FOCP. Heydari et al. [51] developed wavelets method for solving FOCP. Tang et al. [52] implemented an integral fractional pseudo spectral method for solving FOCP. Lotfi et al. [53] proposed a Legendre orthonormal basis combined with the OM and the Gauss quadrature rule for a class of FOCP. Ghomanjani [54] presented a numerical technique for solving the FOCP and fractional Riccati differential equations. Özdemir [55] studied the FOCP for a distributed system in cylindrical coordinates. Interested readers can refer to [56–61] and references therein for works on FOCP. The VO fractional optimal control problems (V-FOCP) can be defined with respect to different definitions of VO fractional derivatives. Research on the solution of V-FOCP is relatively new, and the numerical approximation of these equations is still at an early stage of development. Some recent works on V-FOCP can be found in [62–68].

In the current chapter we focus on a class of V-FOCP with the Caputo fractional derivative in a dynamical system and proposes a new direct computational method based on two different families of basis functions, the GP, to obtain an approximate solution. The problem formulation is as follows:

$$\min \ \mathcal{J}[u] = \int_0^1 \mathcal{F}\left(t, x(t), u(t)\right) dt, \tag{1.1}$$

with the VO fractional dynamical system:

$$ {}_0^C D_t^{\alpha(t)} x(t) = \mathcal{G}\left(t, x(t), u(t)\right), \quad q - 1 < \alpha(t) \le q, \quad t \in [0, 1], \tag{1.2}$$

and the initial conditions:

$$x(0) = a_0, \ x'(0) = a_1, \dots, x^{q-1}(0) = a_{q-1}, \tag{1.3}$$

where q is a positive integer, a_j for $j = 0, 1, \dots, q - 1$ are real constants, $\mathcal{F}$ and $\mathcal{G}$ are continuous functions. The symbol ${}_0^C D_t^{\alpha(t)} x(t) \equiv x^{(\alpha(t))}(t)$ denotes the VO fractional derivative of order $\alpha(t)$ of the Caputo type of $x(t)$ defined as [62–68]:

$$\left({}_0^C D_t^{\alpha(t)} x \right)(t) = \begin{cases} \dfrac{1}{\Gamma(q - \alpha(t))} \displaystyle\int_0^t (t - \tau)^{q - \alpha(t) - 1} \dfrac{d^q x(\tau)}{d\tau^q} \, d\tau, & q - 1 < \alpha(t) < q, \\[4mm] \dfrac{d^q x(t)}{dt^q}, & \alpha(t) = q, \end{cases} \tag{1.4}$$

where $q \in \mathbb{N}$ and $\Gamma(\cdot)$ stands for Gamma function defined for $z > 0$ by

$$\Gamma(z) = \int_0^\infty t^{z-1} e^{-t} dt.$$

One sees that
$$\forall n \in \mathbb{Z}^+ : \ \Gamma(n) = (n - 1)!,$$

where by convention $0! = 1$. It is worth noting that based on the definition of the VO fractional derivative of the Caputo type as above, we have the following useful property:

$$
{}^{C}_{0}D^{\alpha(t)}_{t}t^{m} = \begin{cases} \dfrac{\Gamma(m+1)}{\Gamma(m-\alpha(t)+1)}\,t^{m-\alpha(t)}, & \text{for } m \in \mathbb{N}_0 \text{ and } m \geq q, \\[2mm] 0, & \text{for } m \in \mathbb{N}_0 \text{ and } m < q, \end{cases} \tag{1.5}
$$

where $q - 1 < \alpha(t) \leq q$ and $\mathbb{N}_0 = \mathbb{N} \cup \{0\}$.

The proposed method consists of reducing the V-FOCP in Eq. (1.1) and the VO fractional dynamical system in Eq. (1.2) to a system of nonlinear algebraic equations. For this purpose, the state variable $x(t)$ and the control variable $u(t)$ are expanded by means of two different families of GP with unknown free coefficients (FC) and control parameters (CP). The properties of these basis functions lead to a nonlinear system of algebraic equations that substitute the performance index in Eq. (1.1) and the VO fractional dynamical system in Eq. (1.2) in terms of the unknown FC and CP. Finally, the method of constrained extremum is applied. This technique consists of adjoining the constraints equations derived from the given dynamical system to the performance index by a set of undetermined Lagrange multipliers. The necessary conditions of optimality are derived as a system of algebraic equations in the unknown FC and CP in $x(t)$, $u(t)$ and the Lagrange multipliers using the GP properties.

The remainder of the paper is organized as follows. In Sect. 2, the GP and their properties, including function approximation, convergence analysis and the OM of the VO fractional derivative are obtained. Moreover, the proposed method for solving the V-FOCP in Eq. (1.1) is described. In Sect. 3 numerical examples illustrate the applicability and accuracy of the new method. Finally, Sect. 4 summarises the conclusions of this chapter.

2   Description of the Proposed Method

In this section, we introduce the GP and employ them as a class of basis functions. We present the VO fractional and ordinary differentiation matrices of the GP in the Caputo sense, and we discuss the convergence of the proposed method. Furthermore, we describe the process of solving the V-FOCP (1.1) with the VO fractional dynamical system (1.2) subject to the initial conditions (1.3).

2.1   Constructing Appropriate Basis Functions

Let us start by introducing two families of the GP as follows:

$$\psi_i^{\alpha(t)}(t) = \begin{cases} 1, & i = 1, \\ t^{i-1}, & i = 2, 3, \ldots, \lceil \alpha(t) \rceil, \\ t^{i-1+k_{i-1}}, & i = \lceil \alpha(t) \rceil + 1, \lceil \alpha(t) \rceil + 2, \ldots, m, \end{cases} \tag{2.1}$$

and

$$\varphi_i(t) = \begin{cases} 1, & i = 1, \\ t^{i-1+s_{i-1}}, & i = 2, 3, \ldots, n, \end{cases} \tag{2.2}$$

where k_i and s_i are the CP, and the ceiling function $\lceil \alpha(t) \rceil$ denotes the smallest integer greater than or equal to $\alpha(t)$.

The m and n-sets of such basis functions can be expressed as:

$$\Psi^{\alpha(t)}(t) \triangleq [\psi_1^{\alpha(t)}(t) \ \psi_2^{\alpha(t)}(t) \ \ldots \ \psi_m^{\alpha(t)}(t)]^T, \tag{2.3}$$

and

$$\Phi(t) \triangleq [\varphi_1(t) \ \varphi_2(t) \ \ldots \ \varphi_n(t)]^T. \tag{2.4}$$

Let $\psi_i^{\alpha(t)}(t)$ be Eq. (2.1) and $q - 1 < \alpha(t) \le q$ be a positive function. Then:

$$\frac{d\psi_i^{\alpha(t)}(t)}{dt} = \begin{cases} 0, & i = 1, \\ (i-1)t^{i-2}, & i = 2, 3, \ldots, q, \\ (i-1+k_{i-1})t^{i-2+k_{i-1}}, & i = q+1, q+2, \ldots, m. \end{cases}$$

The first-order derivative of $\Psi^{\alpha(t)}(t)$ may be written as:

$$\frac{d\Psi^{\alpha(t)}(t)}{dt} = D_t^{(1)} \Psi^{\alpha}(t), \tag{2.5}$$

where the $m \times m$ matrix $D_t^{(1)}$ is called the OM of ordinary derivative and its elements can be computed as follows:

$$D_t^{(1)} = \begin{pmatrix} 0 & 0 & 0 & \cdots & 0 & 0 & 0 & \cdots & 0 \\ 0 & \frac{1}{t} & 0 & \cdots & 0 & 0 & 0 & \cdots & 0 \\ 0 & 0 & \frac{2}{t} & \cdots & 0 & 0 & 0 & \cdots & 0 \\ \vdots & \vdots & \vdots & & \vdots & \vdots & \vdots & & \vdots \\ 0 & 0 & 0 & \cdots & \frac{q-1}{t} & 0 & 0 & \cdots & 0 \\ 0 & 0 & 0 & \cdots & 0 & \frac{q+k_q}{t} & 0 & \cdots & 0 \\ 0 & 0 & 0 & \cdots & 0 & 0 & \frac{q+1+k_{q+1}}{t} & \cdots & 0 \\ \vdots & \vdots & \vdots & & \vdots & \vdots & \vdots & & \vdots \\ 0 & 0 & 0 & \cdots & 0 & 0 & 0 & \cdots & \frac{m-1+k_{m-1}}{t} \end{pmatrix}.$$

Generally, the r-order derivative OM of $\Psi^{\alpha(t)}(t)$ can be expressed as follows:

$$\frac{d^r \Psi^{\alpha(t)}(t)}{dt^r} = D_t^{(r)} \Psi^{\alpha(t)}(t). \tag{2.6}$$

The VO fractional derivative of order $\alpha(t)$ in the Caputo type for $\psi_i^{\alpha(t)}(t)$ as follows:

$$
{}_0^C D_t^{\alpha(t)} \psi_i^{\alpha(t)}(t) =
\begin{cases}
\dfrac{\Gamma(i+k_{i-1})}{\Gamma(i-\alpha(t)+k_{i-1})} t^{i-1-\alpha(t)+k_{i-1}}, & i = q+1, q+2, \ldots, m, \\
0, & \text{otherwise.}
\end{cases}
$$

The VO fractional derivative in the Caputo type of the GP vector $\Psi^{\alpha(t)}(t)$ is:

$$
{}_0^C D_t^{\alpha(t)} \Psi^{\alpha(t)}(t) = D_t^{(\alpha(t))} \Psi^{\alpha(t)}(t), \tag{2.7}
$$

where the $m \times m$ matrix $D_t^{(\alpha(t))}$ is called the OM of VO fractional derivative of order $\alpha(t)$ for $\Psi^{\alpha(t)}(t)$ and its elements can be computed as follows:

$$
D_t^{(\alpha(t))} = t^{-\alpha(t)}
\begin{pmatrix}
0\,0\,0\cdots 0 & 0 & 0 & \cdots & 0 \\
0\,0\,0\cdots 0 & 0 & 0 & \cdots & 0 \\
0\,0\,0\cdots 0 & 0 & 0 & \cdots & 0 \\
\vdots\;\vdots\;\vdots\quad\vdots & \vdots & \vdots & & \vdots \\
0\,0\,0\cdots 0 & 0 & 0 & \cdots & 0 \\
0\,0\,0\cdots 0 & \frac{\Gamma(q+1+k_q)}{\Gamma(q+1-\alpha(t)+k_q)} & 0 & \cdots & 0 \\
0\,0\,0\cdots 0 & 0 & \frac{\Gamma(q+2+k_{q+1})}{\Gamma(q+2-\alpha(t)+k_{q+1})} & \cdots & 0 \\
\vdots\;\vdots\quad\vdots & \vdots & \vdots & & \vdots \\
0\,0\,0\cdots 0 & 0 & 0 & \cdots & \frac{\Gamma(m+k_{m-1})}{\Gamma(m-\alpha(t)+k_{m-1})}
\end{pmatrix}.
$$

2.2 Function Approximation

Let $\mathbb{X} = L^2[0, 1]$ and assume that $\Psi^{\alpha(t)}(t)$ is the vector defined in Eq. (2.3), $\mathbb{Y}_m = span\{\psi_1^{\alpha(t)}, \psi_2^{\alpha(t)}, \ldots, \psi_m^{\alpha(t)}\}$ and $\tilde{x}$ is an arbitrary element in $\mathbb{X}$. Since $\mathbb{Y}_m$ is a finite dimensional vector subspace of $\mathbb{X}$, $\tilde{x}$ has a unique best approximation out of $\mathbb{Y}_m$ such as $x_0 \in \mathbb{Y}_m$, with

$$\forall \hat{x} \in \mathbb{Y}_m, \quad \| \tilde{x} - x_0 \|_2 \leq \| \tilde{x} - \hat{x} \|_2 .$$

Since $x_0 \in \mathbb{Y}_m$, there exist the unique coefficients $X^T = [x_1 \ x_2 \ \ldots \ x_m]$, such that

$$x(t) \simeq x_0(t) = X^T \Psi^{\alpha(t)}(t). \tag{2.8}$$

Remark 1 It should be noted that Eq. (2.8) is usually called GP representation of $x(t)$.

Definition 2.1 Let X be a normed space, A is a nonempty subset of X and $x_0 \in X$ then

$$dist(x_0, A) = inf\{\| x_0 - x \|_2; \ x \in A\}.$$

2.3 Convergence Analysis

The following theorem will be useful in subsequent results.

Theorem 2.2 *Let* $f : [0, 1] \longrightarrow \mathbb{R}$ *be a function,* $f \in C^m[0, 1]$ *and* $M = sup_{0 \leq t \leq 1} |f^{(m)}(t)|$. *If there exist* $x_1, x_2, \ldots, x_m \in \mathbb{R}$ *and* $q + 1, q + 2, \ldots, m > 0$ *such that for the function*

$$x(t) \simeq X^T \, \Psi^{\alpha(t)}(t)$$

we have

$$\|f - x\|_2 = dist(f, x),$$

then the error bound is presented as follows:

$$\|f - X^T \, \Psi^{\alpha(t)}(t)\|_2 \leq \frac{M}{m!\sqrt{2m + 1}},$$

where

$$\|f - X^T \, \Psi^{\alpha(t)}(t)\|_2 = inf\{\|f - y\|_2 : \ y \in \mathbb{Y}_m\}.$$

Proof Let $p(t) = \sum_{i=0}^{m-1} \frac{f^{(i)}(0)}{i!} t^i$. In this case:

$$\|f - X^T \, \Psi^{\alpha(t)}(t)\|_2 \leq \|f - p\|_2.$$

We notice that $t^k \in \mathbb{Y}_m$, which implies that $p(t) \in \mathbb{Y}_m$. Now, by the Taylor theorem, for all $0 \leq t \leq 1$, we have:

$$f(t) = p(t) + \frac{f^{(m)}(\eta_t)}{m!} t^m, \quad 0 \leq \eta_t \leq t.$$

Therefore

$$|f(t) - p(t)| \leq \frac{M}{m!} t^m,$$

and

$$\|f - p\|_2 \leq \frac{M}{m!} \left(\int_0^1 t^{2m} \, dt \right)^{\frac{1}{2}}$$

$$= \frac{M}{m!} \left(\frac{1}{2m + 1} \right)^{\frac{1}{2}}$$

$$= \frac{M}{m!\sqrt{2m + 1}}.$$

∎

Remark 2 Note that the same results for $\Phi(t)$ can be investigated.

2.4 Implementation of the Numerical Method

We use the above obtained results to solve the V-FOCP (1.1) with the VO fractional dynamical system (1.2) subject to the initial conditions (1.3). Therefore, we approximate the state and the control variables $x(t)$ and $u(t)$ by $\Psi^{\alpha(t)}(t)$ and $\Phi(t)$, respectively, as follows:

$$x(t) \simeq X^T \Psi^{\alpha(t)}(t),$$
$$u(t) \simeq U^T \Phi(t), \tag{2.9}$$

where

$$X^T \triangleq [x_1 \ x_2 \ \dots \ x_m], \quad U^T \triangleq [u_1 \ u_2 \ \dots \ u_n],$$

are unknowns that called vectors of the FC, to be computed, and $\Psi^{\alpha(t)}(t)$ and $\Phi(t)$ are the vectors defined in Eqs. (2.3) and (2.4), respectively.
We approximate $_0^C D_t^{\alpha(t)} x(t)$ in terms of the OM of VO fractional derivative of GP by Eq. (2.9), as follows:

$$_0^C D_t^{\alpha(t)} x(t) \simeq X^T D_t^{(\alpha(t))} \Psi^{\alpha(t)}(t). \tag{2.10}$$

By substituting Eq. (2.9) into Eq. (1.1), the performance index $\mathcal{J}$ is approximated as:

$$\mathcal{J}[u] = \mathcal{J}[X, U, K, S] = \int_0^1 \mathcal{F}\left(t, X^T \Psi^{\alpha(t)}(t), U^T \Phi(t)\right) dt \triangleq \hat{\mathcal{F}}(X, U, K, S), \tag{2.11}$$

where K and S are unknown CP vectors with elements k_i and s_i, respectively.
    Substituting Eqs. (2.9) and (2.10) into the VO fractional dynamical system (1.2), we have:

$$X^T D^{(\alpha(t))} \Psi^{\alpha(t)}(t) - \mathcal{G}\left(t, X^T \Psi^{\alpha(t)}(t), U^T \Phi(t)\right) \triangleq \hat{\mathcal{G}}\left(t, X, U, K, S\right) \simeq 0. \quad (2.12)$$

By taking the collocation points $t_i = \frac{i}{\hat{m}-1}$ for $i = \lceil\alpha(t)\rceil, \lceil\alpha(t)\rceil + 1, \ldots, \hat{m} - 1$, $\hat{m} = \min(m, n)$, into Eq. (2.12), we obtain the following system of algebraic equations:

$$\Lambda_i \triangleq \hat{\mathcal{G}}\left(t_i, X, U, K, S\right) = 0, \quad i = \lceil\alpha(t)\rceil, \lceil\alpha(t)\rceil + 1, \ldots, \hat{m} - 1. \quad (2.13)$$

In addition, by the initial conditions (1.3) and considering Eq. (2.6), we get the system of equations:

$$\Lambda_i \triangleq X^T D^{(i)} \Psi^{\alpha(t)}(0) - a_i = 0, \quad i = 0, 1, \ldots, \lceil\alpha(t)\rceil - 1. \quad (2.14)$$

Now, assume that

$$\mathcal{J}^*[X, U, K, S, \lambda] = \mathcal{J}[X, U, K, S] + \Lambda\lambda, \quad (2.15)$$

where $\Lambda = [\Lambda_0 \ \Lambda_1 \ \ldots \ \Lambda_{\hat{m}-1}]$ and

$$\lambda = [\lambda_0 \ \lambda_1 \ \ldots \ \lambda_{\hat{m}-1}]^T,$$

is the unknown Lagrange multiplier.

Finally, the necessary conditions for the extremum are given by the following system of nonlinear algebraic equations:

$$\frac{\partial \mathcal{J}^*}{\partial X} = 0, \ \frac{\partial \mathcal{J}^*}{\partial U} = 0, \ \frac{\partial \mathcal{J}^*}{\partial K} = 0, \ \frac{\partial \mathcal{J}^*}{\partial S} = 0, \ \frac{\partial \mathcal{J}^*}{\partial \lambda} = 0. \quad (2.16)$$

The above system of nonlinear algebraic equations can be solved using symbolic software packages. Finally, by determining the FC and CP, we obtain a good approximate solutions for $u(t)$ and $x(t)$ using Eq. (2.9).

The algorithm of the proposed method is as follows:

Algorithm

Input: $m, n, (q - 1) < \alpha(t) \leq q, a_j \ (j = 0, 1, \ldots, q - 1)$ and the functions $\mathcal{F}$ and $\mathcal{G}$

Step 1: Define the basis functions $\psi_i^{\alpha(t)}(t)$ and $\varphi_i(t)$ by Eqs. (2.1) and (2.2)

Step 2: Construct GPs vectors $\Psi^{\alpha(t)}(t)$ and $\Phi(t)$ using Eqs. (2.3) and (2.4)

Step 3: Define the unknown matrices $X^T = [x_i]_{1 \times m}$ and $U^T = [u_i]_{1 \times n}$

Step 4: Compute the variable order fractional and ordinary operational matrices $D_t^{(\alpha(t))}$ and $D_t^{(1)}$ using Eqs. (2.5) and (2.7)

Step 5: Compute the equation $\mathcal{J}[u] \triangleq \hat{\mathcal{F}}(X, U, K, S)$, using (2.11)

Step 6: Compute the system of nonlinear algebraic equations using Eq. (2.12)

Step 7: Determine the free coefficients and control parameters using Eq. (2.16)

Output: The approximate solution: $x(t) \simeq X^T \Psi^{\alpha(t)}(t)$ and $u(t) \simeq U^T \Phi(t)$

---

3 Illustrative Test Problems

In this section, we consider three numerical examples to illustrate the efficiency and reliability of the proposed method for V-FOCP. The maximum absolute errors for these examples in various points $t_i \in [0, 1]$ are obtained as

$$e_x = \max_{t_i \in [0,1]} \left| x(t_i) - X^T \, \Psi^{\alpha(t)}(t_i) \right|, \qquad e_u = \max_{t_i \in [0,1]} \left| u(t_i) - U^T \Phi(t_i) \right|, \qquad e_{\mathcal{J}} = \max_{t_i \in [0,1]} |\mathcal{J} - \mathcal{J}_m|.$$

Example 1 Consider the following V-FOCP:

$$\min \, \mathcal{J}[u] = \int_0^1 \left[\left(x(t) - t^2 \right)^2 + \left(u(t) - t e^{-t} + \frac{1}{2} e^{t^2-t} \right)^2 \right] dt,$$

subject to the nonlinear VO fractional dynamical system:

$$^C_0 D_t^{\alpha(t)} x(t) = e^{x(t)} + 2e^t u(t), \qquad 0 < \alpha(t) \leq 1,$$

with the initial condition $x(0) = 0$.

For the case $\alpha(t) = 1$, the solution $x(t) = t^2$ and $u(t) = t \, e^{-t} - \dfrac{1}{2} e^{t^2-t}$ minimize the performance index $\mathcal{J}$ and its minimum value is 0. To solve this problem, we use the proposed method with $m = 4$ and $n = 5$ for some different variable-orders $\alpha(t)$, namely for

$$\alpha_1(t) = 1.$$
$$\alpha_2(t) = 1 - 0.2 \exp(-100\,t).$$
$$\alpha_3(t) = 0.75 + 0.2 \cos(t^2). \tag{3.1}$$
$$\alpha_4(t) = 1 - 0.1|t - 1| \sin(t^2).$$

Figure 1 shows the behavior of the numerical solutions for the state and control variables $x(t)$ and $u(t)$. The errors obtained by the proposed method for the performance indexes are $e_{\mathcal{J}} = 4.0321E - 7$, $e_{\mathcal{J}} = 5.8576E - 7$, $e_{\mathcal{J}} = 1.5928E - 5$, $e_{\mathcal{J}} = 8.0187E - 6$ for $\alpha_i(t)$, $i = 1, 2, 3, 4$, respectively. We verify that the proposed method provides a good approximate solution with high accuracy for this problem.

Example 2 Consider the following V-FOCP:

$$\min \, \mathcal{J}[u] = \int_0^1 \left[\left(x(t) - t^{\frac{5}{2}} \right)^4 + \left(1 + t^2 \right) \left(u(t) + t^6 - \frac{15\sqrt{\pi}}{8\Gamma\left(\frac{7}{2} - \alpha(t)\right)} t^{\frac{5}{2}-\alpha(t)} \right)^2 \right] dt,$$

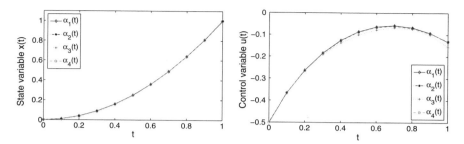

Fig. 1 The behavior of the approximate state variable $x(t)$ (left side) and control variable $u(t)$ (right side) for some different variable-orders $\alpha(t)$ in Example 1 with $m = 4$ and $n = 5$

subject to the nonlinear VO fractional dynamical system:

$$\,^c_0 D^{\alpha(t)}_t x(t) = t\, x^2(t) + u(t), \qquad 1 < \alpha(t) \leq 2,$$

and the initial conditions $x(0) = x'(0) = 0$.

For this problem, the values $x(t) = t^{\frac{5}{2}}$ and $u(t) = \dfrac{15\sqrt{\pi}}{8\Gamma\left(\frac{7}{2} - \alpha(t)\right)} t^{\frac{5}{2} - \alpha(t)} - t^6$ are the minimizing solutions for the state and control variables, respectively, and the performance index $\mathcal{J}$ has the minimum value of 0. We considered $m = 4$ and $n = 7$ in Eq. (2.9) to obtain an approximate solution for the problem. Table 1 lists the absolute errors in the state variable, control variable and performance index with $m = 4$ and $n = 7$ for different variable-orders $\alpha(t)$ as follow:

$$\alpha_1(t) = 1.9.$$
$$\alpha_2(t) = 1.9 + 0.05\, t^2.$$
$$\alpha_3(t) = 1.9 + 0.05 \sin(t^3). \qquad (3.2)$$
$$\alpha_4(t) = 1.9 + 0.05 \cos^2(t^2).$$

Figure 2 depicts the behavior of the numerical solution for the state variable $x(t)$ and control variable $u(t)$ for different variable-orders $\alpha(t)$ with $m = 4$ and $n = 7$. By comparing the numerical results obtained in Table 1, we observed that our results are more accurate than those obtained in [69].

Example 3 Consider the following V-FOCP:

$$\min \mathcal{J}[u] = \int_0^1 \left[e^t \left(x(t) - t^4 + t - 1\right)^2 + (1 + t^2)\left(u(t) + 1 - t + t^4 - \frac{24\, t^{4-\alpha(t)}}{\Gamma(5 - \alpha(t))}\right)^2 \right] dt,$$

Table 1 The maximum absolute errors in the state and control variables and performance index with $m = 4$ and $n = 7$ for Example 2

$\alpha(t)$	e_x	e_u	$e_{\mathcal{J}}$
$\alpha_1(t)$	$6.3417E - 05$	$8.3185E - 04$	$1.9072E - 18$
$\alpha_2(t)$	$7.4045E - 06$	$4.3594E - 04$	$2.1214E - 22$
$\alpha_3(t)$	$4.8272E - 05$	$1.3425E - 04$	$5.0161E - 19$
$\alpha_4(t)$	$5.7403E - 05$	$3.1838E - 04$	$1.5078E - 18$

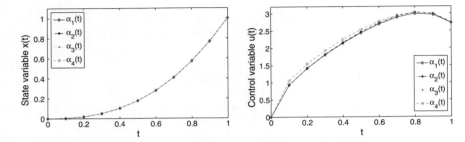

Fig. 2 The behavior of the approximate state variable $x(t)$ and control variable $u(t)$ for different variable-orders $\alpha(t)$ in Example 2 with $m = 4$ and $n = 7$

subject to the VO fractional dynamical system:

$$_0^C D_t^{\alpha(t)} x(t) = x(t) + u(t), \qquad 1 < \alpha(t) \le 2,$$

and the initial conditions $x(0) = 1$, $x'(0) = -1$.

For this problem, $x(t) = 1 - t + t^4$ and $u(t) = \dfrac{24 \, t^{4-\alpha(t)}}{\Gamma(5 - \alpha(t))} - \left(1 - t + t^4\right)$ are the solutions for minimizing the state and control variables, respectively, and the performance index $\mathcal{J}$ has the minimum value of 0. We solve this problem by the proposed method for $m = 4$ and $n = 7$. Table 2 contains the absolute errors in the state variable, control variable and performance index with $m = 4$ and $n = 7$ for the following $\alpha(t)$:

$$\alpha_1(t) = 1.1.$$

$$\alpha_2(t) = 1.1 + 0.00005t^2.$$

$$\alpha_3(t) = 1.1 + 0.00005 \sin(t^3). \tag{3.3}$$

$$\alpha_4(t) = 1.1 + 0.00005 \cos^2(t^2).$$

From Table 2, it is clear that the numerical results are in a good agreement with the exact solutions. Figure 3 demonstrate the behavior of the numerical solutions for the state and control variables $x(t)$ and $u(t)$, respectively. Again, the proposed method provides a good approximate solution for this problem.

Table 2 The maximum absolute errors in the state and control variables and performance index with $m = 4$ and $n = 7$ for Example 3

$\alpha(t)$	e_x	e_u	$e_{\mathcal{J}}$
$\alpha_1(t)$	$5.3334E - 06$	$8.4376E - 04$	$1.5049E - 11$
$\alpha_2(t)$	$4.5258E - 06$	$8.0438E - 04$	$1.1198E - 11$
$\alpha_3(t)$	$8.1231E - 06$	$2.0987E - 04$	$3.7331E - 11$
$\alpha_4(t)$	$4.1544E - 06$	$7.8312E - 04$	$1.0056E - 11$

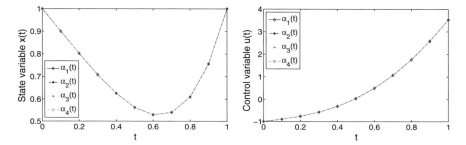

Fig. 3 The behavior of the approximate state variable $x(t)$ and control variable $u(t)$ for some different variable-orders $\alpha(t)$ in Example 3 with $m = 4$ and $n = 7$

4 Conclusions

In this chapter we developed an efficient and accurate technique for solving a class of V-FOCP. The proposed computational method was based on the GP and their OM of VO fractional derivative was in the Caputo sense. Some illustrative test problems are provided to show the efficiency and accuracy of the new method. The numerical solutions demonstrate that a few numbers of GP expansion terms are needed to obtain a good approximation for these problems. Indeed, the results are in a good agreement with the exact solutions. The new OM can be further explored with other types of V-FDE.

References

1. Samko SG, Kilbas AA, Marichev OI (1993) Fractional integrals and derivatives: theory and applications. Breach Science Publishers, London, UK
2. Podlubny I (1999) Fractional differential equations. Academic Press, New York, USA
3. Kilbas AA, Srivastava HM, Trujillo JJ (2006) Theory and applications of fractional differential equations. Elsevier, Amsterdam, The Netherlands
4. Hilfer R (2000) Applications of fractional calculus in physics. World Scientific, Singapore
5. Atanackovic T, Philipović S, Stanković B, Zorica D (2014) Fractional calculus with applications in mechanics: vibrations and diffusion processes. Wiley, London, UK

6. Mainardi F (2010) Fractional calculus and waves in linear viscoelasticity. Imperial College Press and World Scientific, Singapore
7. Zhao M, Wang H (2019) Fast finite difference methods for space-time fractional partial differential equations in three space dimensions with nonlocal boundary conditions. Appl Numer Math 145:411–428
8. Jia J, Wang H (2018) A fast finite difference method for distributed-order space-fractional partial differential equations on convex domains. Comput Math Appl 75(6):2031–2043
9. Sakar MG, Uludag F, Erdogan F (2016) Numerical solution of time-fractional nonlinear PDEs with proportional delays by homotopy perturbation method. Appl Math Model 40(13–14):6639–6649
10. Odibat Z (2019) On the optimal selection of the linear operator and the initial approximation in the application of the homotopy analysis method to nonlinear fractional differential equations. Appl Numer Math 137:203–212
11. Mohammadi F, Cattani C (2018) A generalized fractional-order Legendre wavelet Tau method for solving fractional differential equations. J Comput Appl Math 339:306–316
12. Rahimkhani P, Ordokhani Y, Lima PM (2019) An improved composite collocation method for distributed-order fractional differential equations based on fractional Chelyshkov wavelets. Appl Numer Math 145:1–27
13. Xie J, Wang T, Ren Z, Zhang J, Quan L (2019) Haar wavelet method for approximating the solution of a coupled system of fractional-order integral-differential equations. Math Comput Simulat 163:80–89
14. Heydari MH, Avazzadeh Z, Mahmoudi MR (2019) Chebyshev cardinal wavelets for nonlinear stochastic differential equations driven with variable-order fractional Brownian motion. Chaos Solitons Fract 124:105–124
15. Heydari MH, Hooshmandasl MR, Mohammadi F, Cattani C (2014) Wavelets method for solving systems of nonlinear singular fractional Volterra integro-differential equations. Commun Nonlinear Sci 19(1):37–48
16. Hu Y, Luo Y, Lu Z (2008) Analytical solution of the linear fractional differential equation by Adomian decomposition method. J Comput Appl Math 215(1):220–229
17. Duan JS, Chaolu T, Rach R, Lu L (2013) The Adomian decomposition method with convergence acceleration techniques for nonlinear fractional differential equations. Comput Math Appl 66(5):728–736
18. Yang JY, Huang JF, Liang DM, Tang YF (2014) Numerical solution of fractional diffusion-wave equation based on fractional multistep method. Appl Math Model 38(14):3652–3661
19. Maleki M, Davari A (2019) Fractional retarded differential equations and their numerical solution via a multistep collocation method. Appl Numer Math 143:203–222
20. Jhinga A, Daftardar-Gejji V (2018) A new finite-difference predictor-corrector method for fractional differential equations. Appl Math Comput 336:418–432
21. Diethelm K, Ford NJ, Freed AD (2002) A predictor-corrector approach for the numerical solution of fractional differential equations. Nonlinear Dynam 29(1–4):3–22
22. Bhrawy AH, Zaky MA, Van Gorder RA (2016) A space-time Legendre spectral tau method for the two-sided space-time Caputo fractional diffusion-wave equation. Numer Algorithm 71(1):151–180
23. Abd-Elhameed WM, Youssri YH (2019) Spectral Tau algorithm for certain coupled system of fractional differential equations via generalized fibonacci polynomial sequence. Iranian J Sci Technol A 43(2):543–554
24. Coimbra CFM (2003) Mechanics with variable-order differential operators. Ann Phys 12(11–12):692–703
25. Lorenzo CF, Hartley TT (2002) Variable order and distributed order fractional operators. Nonlinear Dynam 29:57–98
26. Samko SG, Ross B (1993) Intergation and differentiation to a variable fractional order. Integral Trans Special Func 1(4):277–300
27. Samko SG (1995) Fractional integration and differentiation of variable order. Anal Math 21:213–236

28. Meng R, Yin D, Lu S, Xiang G (2019) Variable-order fractional constitutive model for the time-dependent mechanical behavior of polymers across the glass transition. Eur Phys J Plus 134:376

29. Gu Y, Sun H (2020) A meshless method for solving three-dimensional time fractional diffusion equation with variable-order derivatives. Appl Math Model 78:539–549

30. Sheng H, Sun H, Chen YQ, Qiu TS (2011) Synthesis of multifractional Gaussian noises based on variable-order fractional operators. Signal Process 91(7):1645–1650

31. Tseng CC (2006) Design of variable and adaptive fractional order FIR differentiators. Signal Process 86(10):2554–2566

32. Dahaghin MSh, Hassani H (2017) An optimization method based on the generalized polynomials for nonlinear variable-order time fractional diffusion-wave equation. Nonlinear Dynam 88(3):1587–1598

33. Sun HG, Chen W, Chen YQ (2009) Variable-order fractional differential operators in anomalous diffusion modeling. Phys A 388(21):4586–4592

34. Sun HG, Chang A, Zhang Y, Chen W (2019) A review on variable-order fractional differential equations: mathematical foundations, physical models, numerical methods and applications. Fract Calc Appl Anal 22(1):27–59

35. Chen YM, Liu LQ, Li BF, Sun YN (2014) Numerical solution for the variable order linear cable equation with bernstein polynomials. Appl Math Comput 238:329–341

36. Chen YM, Liu LQ, Li X, Sun YN (2014) Numerical solution for the variable order time fractional diffusion equation with Bernstein polynomials. CMES-Comp Model Eng 97(1):81–100

37. Chen YM, Wei YQ, Liu DY, Yu H (2015) Numerical solution for a class of nonlinear variable order fractional differential equations with Legendre wavelets. Appl Math Lett 46:83–88

38. Zayernouri M, Karniadakis GE (2015) Fractional spectral collocation methods for linear and nonlinear variable order FPDEs. J Comput Phys 239:312–338

39. Zhao X, Sun ZZ, Karniadakis GE (2015) Second-order approximations for variable order fractional derivatives: algorithms and applications. J Comput Phys 293:184–200

40. Yaghoobi Sh, Parsa Moghaddam B, Ivaz K (2017) An efficient cubic spline approximation for variable-order fractional differential equations with time delay. Nonlinear Dynam 87(2):815–826

41. Parsa Moghaddam B, Tenreiro Machado JA (2017) Extended algorithms for approximating variable order fractional derivatives with applications. J Sci Comput 71(3):1351–1374

42. Parsa Moghaddam B, Tenreiro Machado JA, Behforooz H (2017) An integro quadratic spline approach for a class of variable-order fractional initial value problems. Chaos Soliton Fract 102:354–360

43. Heydari MH, Hooshmandasl MR, Cattani C, Hariharan G (2017) An optimization wavelet method for multi variable-order fractional differential equations. Fund Inform 151(1–4):255–273

44. Tayebi A, Shekari Y, Heydari MH (2017) A meshless method for solving two-dimensional variable-order time fractional advection? diffusion equation. J Comput Phys 340:655–669

45. Hajipour M, Jajarmi A, Baleanu D, Sun HG (2019) On an accurate discretization of a variable-order fractional reaction-diffusion equation. Commun Nonlinear Sci 69:119–133

46. Xiang M, Zhang B, Yang D (2019) Multiplicity results for variable-order fractional Laplacian equations with variable growth. Nonlinear Anal Theor 178:190–204

47. Liu J, Li X, Hu X (2019) A RBF-based differential quadrature method for solving two-dimensional variable-order time fractional advection-diffusion equation. J Comput Phys 384:222–238

48. Hassani H, Avazzadeh Z, Tenreiro Machado JA (2019) Numerical approach for solving variable-order space-time fractional telegraph equation using transcendental Bernstein series. Eng Comput (2019). https://doi.org/10.1007/s00366-019-00736-x

49. Yavari M, Nazemi A (2019) An efficient numerical scheme for solving fractional infinite-horizon optimal control problems. ISAT. https://doi.org/10.1016/j.isatra.2019.04.016

50. Salati AB, Shamsi M, Torres DFM (2019) Direct transcription methods based on fractional integral approximation formulas for solving nonlinear fractional optimal control problems. Commun Nonlinear Sci 67:334–350

51. Heydari MH, Hooshmandasl MR, Maalek Ghaini FM, Cattani C (2016) Wavelets method for solving fractional optimal control problems. Appl Math Comput 286:139–154

52. Tanga X, Liu Z, Wang X (2015) Integral fractional pseudospectral methods for solving fractional optimal control problems. Automatica 62:304–311

53. Lotfi A, Yousefi SA, Dehghanb M (2013) Numerical solution of a class of fractional optimal control problems via the Legendre orthonormal basis combined with the operational matrix and the Gauss quadrature rule. J Comput Appl Math 250:143–160

54. Ghomanjani F (2016) A numerical technique for solving fractional optimal control problems and fractional riccati differential equations. J Egypt Math Soc 24(4):638–643

55. Özdemir N, Karadeniz D, Iskender BB (2009) Fractional optimal control problem of a distributed system in cylindrical coordinates. Phys Lett A 373(2):221–226

56. Pan B, Ma Y, Ni Y (2019) A new fractional homotopy method for solving nonlinear optimal control problems. Acta Astronaut 161:12–23

57. Kamocki R (2014) On the existence of optimal solutions to fractional optimal control problems. Appl Math Comput 235:94–104

58. Rabiei K, Ordokhani Y (2018) Boubaker hybrid functions and their application to solve fractional optimal control and fractional variational problems. Appl Math-Czech 63(5):541–567

59. Rosa S, Torres DFM (2018) Optimal control of a fractional order epidemic model with application to human respiratory syncytial virus infection. Chaos Soliton Fract 117:142–149

60. Kubyshkin VA, Postnov SS (2018) Time-optimal boundary control for systems defined by a fractional order diffusion equation. Automat Rem Contr 79(5):884–896

61. Jahanshahi S, Torres DFM (2017) A simple accurate method for solving fractional variational and optimal control problems. J Optimiz Theory App 174(1):156–175

62. Zaky MA, Tenreiro Machado JA (2017) On the formulation and numerical simulation of distributed-order fractional optimal control problems. Commun Nonlinear Sci 52:177–189

63. Heydari MH (2018) A new direct method based on the Chebyshev cardinal functions for variable-order fractional optimal control problems. J Franklin Inst 355(12):4970–4995

64. Hassani H, Tenreiro Machado JA, Naraghirad E, Generalized shifted Chebyshev polynomials for fractional optimal control problems. Commun Nonlinear Sci Numer Simul 75:50–61

65. Hassani H, Avazzadeh Z (2019) Transcendental Bernstein series for solving nonlinear variable order fractional optimal control problems. Appl Math Comput. https://doi.org/10.1016/j.amc.2019.124563

66. Hassani H, Avazzadeh Z, Tenreiro Machado JA (2019) Solving two-dimensional variable-order fractional optimal control problems with transcendental Bernstein series. J Comput Nonlin Dyn 14(6):061001–11

67. Zaky MA (2018) A Legendre collocation method for distributed-order fractional optimal control problems. Nonlinear Dynam 91:2667–2681

68. Mohammadi F, Hassani H (2019) Numerical solution of two-dimensional variable-order fractional optimal control problem by generalized polynomial basis. J Optim Theory Appl 180(2):536–555

69. Heydari MH, Hooshmandasl MR, Shakiba A, Cattani C (2016) An efficient computational method based on the hat functions for solving fractional optimal control problems. Tbilisi Math J 9:143–157

Solving Nonlinear Variable-Order Time Fractional Convection-Diffusion Equation with Generalized Polynomials

H. Hassani, J. A. Tenreiro Machado, Z. Avazzadeh, and E. Naraghirad

Abstract This chapter presents a new nonlinear variable-order (VO) time fractional convection-diffusion equation (NV-TFCDE). The model generalizes the standard fixed-order nonlinear time fractional convection-diffusion equation. The VO time fractional derivative is described in the Caputo type leading to an optimization method for the NV-TFCDE. The proposed approach is based on a new class of basis functions, namely the generalized polynomials (GP). The solution of the problem under consideration is expanded in terms of the GP with unknown free coefficients (FC) and control parameters (CP). A new VO time fractional operational matrix (V-TFOM) for the GP transforms the problem into a system of nonlinear algebraic equations with unknown FC and CP leading to the approximate solution. Several numerical examples show that the proposed method is efficient.

Keywords Variable-order time fractional convection-diffusion equation ·
Generalized polynomials · Variable-order time fractional operational matrix ·
Optimization method · Free coefficients · Control parameters

H. Hassani (✉)
Faculty of Mathematics and Statistics, Ton Duc Thang University, Ho Chi Minh City, Vietnam
e-mail: hosseinhassani@tdtu.edu.vn

J. A. Tenreiro Machado
Department of Electrical Engineering, Institute of Engineering, Polytechnic of Porto, Porto, Portugal
e-mail: jtm@isep.ipp.pt

R. Dr. António Bernardino de Almeida, 4249-015 Porto 431, Portugal

Z. Avazzadeh
Department of Applied Mathematics, Xi'an Jiaotong-Liverpool University, Suzhou 215123, Jiangsu, China
e-mail: zakieh.avazzadeh@xjtlu.edu.cn

E. Naraghirad
Department of Mathematics, Yasouj University, Yasouj, Iran
e-mail: esnaraghirad@yu.ac.ir

1 Introduction

Variable-order fractional derivatives (V-FD) generalize the constant-order fractional derivatives and can be a function of the space and/or time. The V-FD can be applied in various areas of science and engineering. We find applications of V-FD in physics, mechanics for modeling of linear and nonlinear viscoelasticity oscillators, signal processing, optimization and control [1–7]. Since the equations including V-FD are complex and difficult to solve analytically, it is important to consider their approximate solutions. Recently, a few methods were proposed to obtain the numerical solutions of variable-order (VO) fractional differential equations (V-FDE). Shen et al. [8] presented a numerical technique for the VO time fractional diffusion equation and investigated its stability, convergence and solvability. Yang et al. [9] developed a VO fractional operator for the anomalous diffusion equations and analyzed its properties in term of the Laplace and Fourier transforms. Bhrawy et al. [10] presented an accurate spectral collocation method for solving one-and two-dimensional VO fractional nonlinear cable equations. Zayernouri et al. [11] employed fractional spectral collocation schemes for linear and nonlinear V-FDE. These researchers solved several V-FDE, including the time and space-fractional advection, time and space-fractional advection-diffusion and the space-fractional Burgers' equations. Several other methods have been proposed to solve numerically V-FDE, see e.g. [12–34] and references therein.

The nonlinear VO time fractional convection-diffusion equation (NV-TFCDE) is a nonlinear fractional partial differential equation that is obtained from the nonlinear time fractional convection-diffusion model by replacing the standard fractional time derivative terms with V-FD:

$$
{}_0^C D_t^{\alpha(x,t)} u(x, t) = u_{xx}(x, t) - \eta\, u_x(x, t) + H(u(x, t)) + f(x, t),
$$
$$
(x, t) \in \Omega = [0, 1] \times [0, 1], \tag{1.1}
$$

subject to the initial and boundary conditions:

$$
u(x, 0) = g(x),
$$
$$
u(0, t) = h_1(t),\ u(1, t) = h_2(t), \tag{1.2}
$$

where $H(u(x, t))$ represents a nonlinear operator, η denotes a constant parameter, $0 < \alpha(x, t) \le 1$, and $f(x, t)$, $g(x)$, $h_1(t)$ and $h_2(t)$ are given functions.

Here, ${}_0^C D_t^{\alpha(x,t)} u(x, t)$ denotes the V-FD in the Caputo sense and is defined as [13]:

$$
{}_0^C D_t^{\alpha(x,t)} u(x, t) = \frac{1}{\Gamma\left(1 - \alpha(x, t)\right)} \int_0^t (t - s)^{-\alpha(x,t)}\, \frac{\partial u(x, s)}{\partial s}\, ds, \qquad t > 0. \tag{1.3}
$$

So, based on the definition of the V-FD in the Caputo sense, we have the following useful property:

$$\begin{matrix} C \\ 0 \end{matrix} D_t^{\vartheta(x,t)} t^m = \begin{cases} \dfrac{\Gamma(m+1)}{\Gamma(m - \vartheta(x,t) + 1)} \, t^{m - \vartheta(x,t)}, & n \le m \in \mathbb{N}, \\ 0, & \text{otherwise}, \end{cases} \qquad (1.4)$$

where $n - 1 < \vartheta(x,t) \le n$.

In this chapter, we propose an optimization method based on new polynomials, namely the GP, to solve equation in (1.1). We first obtain a new VO time fractional derivative operational matrix (V-TFDOM) for the GP in order to obtain the approximate solution of the problem. The solution is expanded in terms of the GP with unknown free coefficients (FC) and control parameters (CP). The residual function and its error in 2-norm are employed for converting the problem to an optimization one and then to choose the optimal values of the unknown FC and CP. Finally, the method of constrained extremum is applied. The strategy consists of adjoining the constraint equations, obtained from the given initial and boundary conditions, to the object function, derived from the residual functions by a set of undetermined Lagrange multipliers. As a result, the necessary conditions of optimality are expressed as a system of nonlinear algebraic equations with unknown FC, CP and Lagrange multipliers. The efficiency and accuracy of the suggested method are illustrated by numerical examples.

This chapter is organized as follows. Section 2, describes the GP and their properties as an effective tool to approximate solution of Eq. (1.1). Section 3, presents several numerical examples that analyse the accuracy and efficiency of method. Finally, Sect. 4, summaries the main conclusions.

2   Description of the Proposed Method

In this section, we introduce the GP and use them as a new class of basis functions. In order to solve the Eq. (1.1) subject to the initial and boundary conditions (1.2), we present a procedure to derive the V-TFDOM in the Caputo type.

2.1   The GP and the Operational Matrices of GP

Let us assume that the GP of degrees m_1 and m_2 are denoted by $\varphi_i(x)$ and $\psi_j(t)$ and that they are generated using the formulae:

$$\varphi_i(x) = \begin{cases} x^{i-1}, & i = 1, 2, \\ x^{i-1+k_{i-1}}, & i = 3, 4, \ldots, m_1 + 1, \end{cases} \qquad (2.1)$$

and

$$\psi_j(t) = \begin{cases} t^{j-1}, & j = 1, \\ t^{j-1+s_{j-1}}, & j = 2, 3, 4, \ldots, m_2 + 1, \end{cases} \tag{2.2}$$

where the symbols k_i and s_j stand for the CP.
The $(m_1 + 1)$-set and $(m_2 + 1)$-set of these basis functions can be expressed as:

$$\Phi_{m_1}(x) \triangleq [\varphi_1(x) \ \varphi_2(x) \ \ldots \ \varphi_{m_1+1}(x)]^T, \tag{2.3}$$

and

$$\Psi_{m_2}(t) \triangleq [\psi_1(t) \ \psi_2(t) \ \ldots \ \psi_{m_2+1}(t)]^T. \tag{2.4}$$

Let $\varphi_i(x)$ be defined as in Eq. (2.1). Then, we have:

$$\frac{d\varphi_i(x)}{dx} = \begin{cases} 0, & i = 1, \\ 1, & i = 2, \\ (i - 1 + k_{i-1}) x^{i-2+k_{i-1}}, & i = 3, 4, \ldots, m_1 + 1, \end{cases}$$

and

$$\frac{d^2\varphi_i(x)}{dx^2} = \begin{cases} 0, & i = 1, 2, \\ (i - 1 + k_{i-1}) (i - 2 + k_{i-1}) x^{i-3+k_{i-1}}, & i = 3, 4, \ldots, m_1 + 1. \end{cases}$$

Eq. (2.3), enables one to write

$$\frac{d\Phi_{m_1}(x)}{dx} = D_x^{(1)} \Phi_{m_1}(x), \qquad \frac{d^2\Phi_{m_1}(x)}{dx^2} = D_x^{(2)} \Phi_{m_1}(x), \tag{2.5}$$

where the dimensions of the square matrices $D_x^{(1)}$ and $D_x^{(2)}$ are $(m_1 + 1) \times (m_1 + 1)$. These matrices are the operational matrices of derivatives $\Phi_{m_1}(x)$ and

$$D_x^{(1)} = \begin{pmatrix} 0 & 0 & 0 & \cdots & 0 \\ 0 & \frac{1}{x} & 0 & \cdots & 0 \\ 0 & 0 & \frac{2+k_2}{x} & \cdots & 0 \\ \vdots & \vdots & \vdots & & \vdots \\ 0 & 0 & 0 & \cdots & \frac{m_1+k_{m_1}}{x} \end{pmatrix}, \quad D_x^{(2)} = \begin{pmatrix} 0 & 0 & 0 & \cdots & 0 \\ 0 & 0 & 0 & \cdots & 0 \\ 0 & 0 & \frac{(2+k_2)(1+k_2)}{x^2} & \cdots & 0 \\ \vdots & \vdots & \vdots & & \vdots \\ 0 & 0 & 0 & \cdots & \frac{(m_1+k_{m_1})(m_1-1+k_{m_1})}{x^2} \end{pmatrix}.$$

Let us consider $\psi_j(t)$ defined in Eq. (2.2) and $0 < \alpha(x, t) \leq 1$ to be a positive function. One can express the V-FD of orders $\alpha(x, t)$ of $\psi_j(t)$ in the Caputo type as follows:

$$
{}^C_0 D^{\alpha(x,t)}_t \psi_j(t) = \begin{cases} \dfrac{\Gamma\left(j+s_{j-1}\right)}{\Gamma\left(j-\alpha(x,t)+s_{j-1}\right)} t^{j-1-\alpha(x,t)+s_{j-1}}, & j=2,3,\ldots,m_2+1, \\ 0, & j=1. \end{cases}
$$

The $\alpha(x,t)$-order derivative of $\Psi_{m_2}(t)$ may be written as:

$$
{}^C_0 D^{\alpha(x,t)}_t \Psi_{m_2}(t) = D^{(\alpha(x,t))}_t \Psi_{m_2}(t), \tag{2.6}
$$

where the $(m_2+1) \times (m_2+1)$ matrix $D^{(\alpha(x,t))}_t$ is called the V-TFDOM of orders $\alpha(x,t)$ for $\Psi_{m_2}(t)$ and its elements are computed as follows:

$$
D^{(\alpha(x,t))}_t = t^{-\alpha(x,t)} \begin{pmatrix} 0 & 0 & 0 & \cdots & 0 \\ 0 & \frac{\Gamma(2+s_1)}{\Gamma(2-\alpha(x,t)+s_1)} & 0 & \cdots & 0 \\ 0 & 0 & \frac{\Gamma(3+s_2)}{\Gamma(3-\alpha(x,t)+s_2)} & \cdots & 0 \\ 0 & 0 & 0 & \cdots & 0 \\ \vdots & \vdots & \vdots & & \vdots \\ 0 & 0 & 0 & \cdots & \frac{\Gamma\left(m_2+1+s_{m_2}\right)}{\Gamma\left(m_2+1-\alpha(x,t)+s_{m_2}\right)} \end{pmatrix}.
$$

2.2   Function Approximation

Let $X = L^2[0,1] \times [0,1]$ and $Y = \langle x^{\beta_i} t^{\gamma_j};\ 0 \le i \le m_1,\ 0 \le j \le m_2 \rangle$. Then, Y is a finite dimensional vector subspace of X ($dim Y \le (m_1+1)(m_2+1) < \infty$) and each $\tilde{u} = \tilde{u}(x,t) \in X$ has a unique best approximation $u_0 = u_0(x,t) \in Y$, i.e.,

$$
\forall \hat{u} \in Y, \quad \| \tilde{u} - u_0 \|_2 \le \| \tilde{u} - \hat{u} \|_2 .
$$

Since $u_0 \in Y$, there exist unique coefficients $u_{ij} \in \mathbb{R}$ such that the dependent variable $u_0(x,t)$ may be expanded in terms of the GP as

$$
u_0(x,t) = \sum_{i=0}^{m_1} \sum_{j=0}^{m_2} u_{ij} x^{\beta_i} t^{\gamma_j} = \Phi_{m_1}(x)^T U \Psi_{m_2}(t), \tag{2.7}
$$

where $\Phi_{m_1}(x)^T$ and $\Psi_{m_2}(t)$ defined in Eqs. (2.3) and (2.4), respectively.

Definition 2.1 Let X be a normed space, A is a nonempty subset of X and $x_0 \in X$. Then

$$
dist(x_0, A) = inf \{ \| x_0 - x \|_2;\ x \in A \} .
$$

Theorem 2.2 *If X is a normed linear space, and Y_1 and Y_2 are two nonempty subsets of X such that $Y_1 \subseteq Y_2$ and $x \in X$, then $dist(x, Y_2) \le dist(x, Y_1)$.*

Theorem 2.3 *If $\beta_0, \ldots, \beta_{n_1} > 0, \gamma_0, \ldots, \gamma_{n_2} > 0, m_1 \leq n_1, m_2 \leq n_2, X = L^2[0, 1] \times$*
$[0, 1], Y_1 = \langle x^{\beta_i} t^{\gamma_j}; \ 0 \leq i \leq m_1, \ 0 \leq j \leq m_2 \rangle, Y_2 = \langle x^{\beta_i} t^{\gamma_j}; \ 0 \leq i \leq n_1, \ 0 \leq j \leq n_2 \rangle, \tilde{u} \in$
X and u_i is the best approximation of $\tilde{u} \in Y_i$ $(i = 1, 2)$, then $\| \tilde{u} - u_2 \|_2 \leq \| \tilde{u} -$
$u_1 \|_2$.

Proof By using Theorem 2.2, the proof is straightforward.

2.3 Numerical Algorithms

This section applies the V-TFDOM and ordinary derivatives for solving Eq. (1.1) subject to initial and boundary conditions (1.2). For this purpose, we approximate the function $u(x, t)$ with GP as:

$$u(x, t) \simeq \Phi_{m_1}(x)^T U \Psi_{m_2}(t), \tag{2.8}$$

where $U = [u_{ij}]$ is an $(m_1 + 1) \times (m_2 + 1)$ unknown matrix (called the vector of free coefficients), to be computed, and $\Phi_{m_1}(x)$ and $\Psi_{m_2}(t)$ are the vectors defined in Eqs. (2.3) and (2.4), respectively.

We approximate ${}^C_0 D_t^{\alpha(x,t)} u(x, t), u_x(x, t)$ and $u_{xx}(x, t)$ in terms of the V-TFDOM and ordinary derivatives of GP as follows:

$$\begin{aligned} {}^C_0 D_t^{\alpha(x,t)} u(x, t) &\simeq \Phi_{m_1}(x)^T U D_t^{(\alpha(x,t))} \Psi_{m_2}(t), \\ u_x(x, t) &\simeq \Phi_{m_1}(x)^T \left(D_x^{(1)}\right)^T U \Psi_{m_2}(t), \\ u_{xx}(x, t) &\simeq \Phi_{m_1}(x)^T \left(D_x^{(2)}\right)^T U \Psi_{m_2}(t). \end{aligned} \tag{2.9}$$

Applying Eq. (2.8) into the initial and boundary conditions (1.2) yields:

$$\begin{aligned} \Lambda_1(x) &\triangleq \Phi_{m_1}(x)^T U \Psi_{m_2}(0) - g(x), \\ \Lambda_2(t) &\triangleq \Phi_{m_1}(0)^T U \Psi_{m_2}(t) - h_1(t), \quad \Lambda_3(t) \triangleq \Phi_{m_1}(1)^T U \Psi_{m_2}(t) - h_2(t). \end{aligned} \tag{2.10}$$

Actually, we are going to obtain the minimum square error of equality by replacing suitable values of u_{ij} $(1 \leq i \leq m_1 + 1, 1 \leq j \leq m_2 + 1), k_i$ $(2 \leq i \leq m_1)$ and s_j $(1 \leq j \leq m_2)$ in Eq. (1.1). Finally, if we employ Eqs. (2.8) and (2.9) into the NV-TFCDE (1.1), then the residual function $\mathcal{R}(x, t)$ can be written as:

$$\Phi_{m_1}(x)^T \left[U D_t^{(\alpha(x,t))} - \left(D_x^{(2)}\right)^T U + \eta \left(D_x^{(1)}\right)^T U \right] \Psi_{m_2}(t) - H(u(x, t)) - f(x, t) \triangleq \mathcal{R}(x, t) \tag{2.11}$$

and the following 2-norm of the residual function:

$$\mathcal{M}(U; k_2, k_3, \ldots, k_{m_1}; s_1, s_2, \ldots, s_{m_2}) = \int_0^1 \int_0^1 \mathcal{R}(x, t)^2 dx dt. \qquad (2.12)$$

Therefore, to obtain the optimal value for the unknown matrix U and the CP k_i and s_j, we consider the following optimization problem:

$$\min \mathcal{J}[U; k_2, k_3, \ldots, k_{m_1}; s_1, s_2, \ldots, s_{m_2}] = \mathcal{M}(U; k_2, k_3, \ldots, k_{m_1}; s_1, s_2, \ldots, s_{m_2}), \quad (2.13)$$

subject to

$$\begin{cases} \Lambda_1 \left(\dfrac{i-1}{m_1} \right) = 0, & i = 2, 3, \ldots, m_1, \\[2mm] \Lambda_2 \left(\dfrac{j-1}{m_2} \right) = 0, & j = 1, 2, \ldots, m_2 + 1, \\[2mm] \Lambda_3 \left(\dfrac{j-1}{m_2} \right) = 0, & j = 1, 2, \ldots, m_2 + 1, \end{cases} \qquad (2.14)$$

where $\mathcal{J}$ is the objective function to be minimized over U and CP k_i ($i = 2, 3, \ldots, m_1$) and s_j ($j = 1, 2, \ldots, m_2$) with equality constraints Λ_i for $i = 1, 2, 3$. We use the Lagrange multiplier method in order to solve the above minimization problem as:

$$\mathcal{J}^*[U; k_2, k_3, \ldots, k_{m_1}; s_1, s_2, \ldots, s_{m_2}; \lambda] = \mathcal{J}[U; k_2, k_3, \ldots, k_{m_1}; s_1, s_2, \ldots, s_{m_2}] + \lambda \Lambda,$$
$$(2.15)$$

where $\lambda = [\lambda_1, \lambda_2, \ldots, \lambda_{m_1 + 2m_2 + 1}]$ is the unknown Lagrange multiplier and Λ is a known column vector, so that its entries are the equality constraints expressed in Eq. (2.14). The necessary conditions for the extremum of the NV-TFCDE (1.1) subject to initial and boundary conditions (1.2) can be written as:

$$\begin{cases} \dfrac{\partial \mathcal{J}^*}{\partial U} = 0, & \dfrac{\partial \mathcal{J}^*}{\partial \lambda} = 0, \\[2mm] \dfrac{\partial \mathcal{J}^*}{\partial k_i} = 0, & i = 2, 3, \ldots, m_1, \\[2mm] \dfrac{\partial \mathcal{J}^*}{\partial s_j} = 0, & j = 1, 2, \ldots, m_2. \end{cases} \qquad (2.16)$$

We can solve the above system of nonlinear algebraic equations and the unknown FC and CP may be determined using symbolic software packages.

The process can be described briefly in the following algorithm to obtain the approximate solution for the problem (1.1).

Algorithm

Input: η; $0 < \alpha(x, t) \le 1$ and the functions $g(x)$, $h_1(t)$, $h_2(t)$, $H(u(x, t))$ and $f(x, t)$

Step 1: Define the basis functions $\varphi_i(x)$ and $\psi_j(t)$ by Eqs. (2.1) and (2.2)

Step 2: Construct GPs vectors $\Phi_{m_1}(x)$ and $\Psi_{m_2}(t)$ using Eqs. (2.3) and (2.4)

Step 3: Define the unknown matrix $U = [u_{ij}]_{(m_1+1) \times (m_2+1)}$

Step 4: Compute the operational matrices $D_x^{(1)}$, $D_x^{(2)}$ and $D_t^{(\alpha(x,t))}$ using Eqs. (2.5) and (2.6)

Step 5: Compute the functions $\Lambda_i(x)$ for $i = 1$ and $\Lambda_j(t)$ for $j = 2, 3$ using Eq. (2.10)

Step 6: Compute the residual function $\mathcal{R}(x, t)$ using Eq. (2.11)

Step 7: Compute the 2-norm of the residual function $\mathcal{R}(x, y)$ using Eq. (2.12)

Step 8: Compute the objective function $\mathcal{J}[U; k_2, k_3, \ldots, k_{m_1}; s_1, s_2, \ldots, s_{m_2}]$ using Eq. (2.13)

Step 9: Minimize the objective function $\mathcal{J}^*[U; k_2, k_3, \ldots, k_{m_1}; s_1, s_2, \ldots, s_{m_2}; \lambda]$ subject to the constraints (2.14)

Output: The approximate solution is $u(x, t) \simeq \Phi_{m_1}(x)^T U \Psi_{m_2}(t)$

---

3 Numerical Examples

In order to illustrate the accuracy and efficiency of the proposed method we consider four examples. The computations are performed by MAPLE 17 with 20 decimal digits on a X64-based PC with Intel (R) CPU Core i5, 3.00 GHz with 4.0 GB of RAM. The absolute errors of the proposed method for these examples in various points $(x, t) \in \Omega$ are calculated as:

$$|e(x_i, t_i)| = \left| \Phi_{m_1}(x_i)^T U \Psi_{m_2}(t_i) - u(x_i, t_i) \right|, \qquad (x_i, t_i) \in \Omega = [0, 1] \times [0, 1]$$

Example 1 Consider the NV-TFCDE (1.1) subject to the initial and boundary conditions:

$$u(x, 0) = x^2,$$

$$u(0, t) = t^2, \quad u(1, t) = 1 + t^2.$$

The nonlinear operator H and the function $f(x, t)$ are given by:

$$H(u(x, t)) = u(x, t)u_x(x, t) - 2xu(x, t), \qquad f(x, t) = \frac{2}{\Gamma(3 - \alpha(x, t))} t^{2-\alpha(x,t)} - 2 + 2\eta x.$$

The exact solution in this example is $u(x, t) = t^2 + x^2$. We solve this problem by the proposed method with $m_1 = 3$, $m_2 = 3$, $\eta = 1$ and $\alpha(x, t) = 0.8 \pm 0.2e^{-x} \sin(t)$. The absolute errors $|e(x_i, t_i)|$ of the approximate solutions in different points $(x, t) \in \Omega$ are listed in Table 1. The plots of the approximate solution and absolute error for $\alpha(x, t) = 0.8 + 0.2e^{-x} \sin(t)$ are shown in Fig. 1. The CP and FC for this problem with $\alpha(x, t) = 0.8 + 0.2e^{-x} \sin(t)$ are obtained as follows:

Table 1 The absolute errors $|e(x_i, t_i)|$ of the approximate solutions in different points for Example 1

(x, t)	$\alpha(x, t) = 0.8 + 0.2e^{-x} \sin(t)$	$\alpha(x, t) = 0.8 - 0.2e^{-x} \sin(t)$
$(0.1, 0.1)$	1.0883E-19	4.8161E-20
$(0.2, 0.2)$	1.8829E-19	7.6328E-20
$(0.3, 0.3)$	2.1415E-19	6.9125E-20
$(0.4, 0.4)$	1.7857E-19	3.6385E-20
$(0.5, 0.5)$	9.2215E-20	2.0439E-21
$(0.6, 0.6)$	1.8045E-21	2.2341E-20
$(0.7, 0.7)$	3.8971E-20	8.5079E-21
$(0.8, 0.8)$	2.3911E-20	3.3310E-20
$(0.9, 0.9)$	1.2957E-19	6.2701E-20

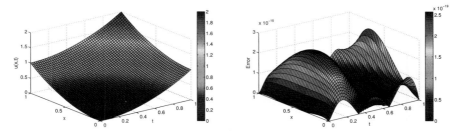

Fig. 1 The approximate solution (left) and absolute error (right) with $m_1 = m_2 = 3$ and $\alpha(x, t) = 0.8 + 0.2e^{-x} \sin(t)$ for Example 1

$$s_1 = 1, \quad s_2 = 0.6998986026, \quad s_3 = 0.7479170701, \quad k_2 = 0, \quad k_3 = 0.3681492424,$$

$$u_{11} = 0, \quad u_{12} = 1, \quad u_{13} = 0, \quad u_{14} = 0, \quad u_{21} = 0, \quad u_{22} = 0, \quad u_{23} = 0, \quad u_{24} = 0,$$

$$u_{31} = 1, \quad u_{32} = 0, \quad u_{33} = 0, \quad u_{34} = 0, \quad u_{41} = 0, \quad u_{42} = 0, \quad u_{43} = 0, \quad u_{44} = 0.$$

The runtimes of the proposed algorithm with $\alpha(x, t) = 0.8 + 0.2e^{-x} \sin(t)$ and $\alpha(x, t) = 0.8 - 0.2e^{-x} \sin(t)$ are about 289 and 276 s, respectively. From Table 1 and Fig. 1 one can see that the approximate solution is considerably accurate.

Example 2 Consider the NV-TFCDE (1.1) subject to the initial and boundary conditions:

$$u(x, 0) = 0,$$

$$u(0, t) = t^3, \ u(1, t) = t^3 e^{-1}.$$

The nonlinear operator H and the function $f(x, t)$ are:

$$H(u(x, t)) = u^2(x, t), \qquad f(x, t) = \left(\frac{6}{\Gamma(4 - \alpha(x, t))} t^{3 - \alpha(x, t)} - t^3 - \eta t^3 - t^6 e^{-x} \right) e^{-x}.$$

Table 2 The absolute errors $|e(x_i, t_i)|$ of the approximate solutions in different points for Example 2

(x, t)	$\alpha(x, t) =$ $1 - 0.04 \cos^2(t) \sin^2(x)$	$\alpha(x, t) =$ $1 - 0.8 \cos^2(t) \sin^2(x)$
$(0.1, 0.1)$	1.1941E-08	1.2783E-08
$(0.2, 0.2)$	1.3053E-07	1.7688E-08
$(0.3, 0.3)$	5.2217E-07	3.6334E-07
$(0.4, 0.4)$	6.1507E-07	8.8843E-07
$(0.5, 0.5)$	4.5640E-07	7.0861E-07
$(0.6, 0.6)$	1.7737E-06	3.3068E-07
$(0.7, 0.7)$	4.8746E-07	1.9417E-06
$(0.8, 0.8)$	3.9024E-06	5.4475E-06
$(0.9, 0.9)$	4.9387E-06	4.5816E-06

The exact solution of this problem is $u(x, t) = t^3 e^{-x}$. We take $m_1 = 5$, $m_2 = 3$, $\eta = 2$ for the VO given by $\alpha(x, t) = 1 - 0.04 \cos^2(t) \sin^2(x)$ and $\alpha(x, t) = 1 - 0.8 \cos^2(t) \sin^2(x)$. The absolute errors $|e(x_i, t_i)|$ of the approximate solutions in different points $(x, t) \in \Omega$ are listed in Table 2. The approximate solution and the absolute error for $\alpha(x, t) = 1 - 0.04 \cos^2(t) \sin^2(x)$ are depicted in Fig. 2. The CP and FC for $\alpha(x, t) = 1 - 0.04 \cos^2(t) \sin^2(x)$ are obtained as follows:

$s_1 = 2.0016571600$, $\quad s_2 = 0.9997969694$, $\quad s_3 = 0.00577173701$, $\quad k_2 = -0.0469392651$,

$k_3 = 0.5777638782$, $\quad k_4 = 0.6631007765$, $\quad k_5 = 0.7705042416$, $\quad u_{11} = 0$, $\quad u_{12} = 0.1564682638$,

$u_{13} = 0.8582043328$, $\quad u_{14} = -0.01467260$, $\quad u_{21} = 0$, $\quad u_{22} = -0.4195707249$, $\quad u_{23} = -0.6942265615$,

$u_{24} = 0.1124912440$, $\quad u_{31} = 0$, $\quad u_{32} = 0.5560455654$, $\quad u_{33} = 0.0923069103$, $\quad u_{34} = -0.1990660126$,

$u_{41} = 0$, $\quad u_{42} = -0.1148614174$, $\quad u_{43} = -0.0601669819$, $\quad u_{44} = 0.0384085542$, $\quad u_{51} = 0$,

$u_{52} = -0.0588283545$, $\quad u_{53} = 0.0586037560$, $\quad u_{54} = 0.0737490821$, $\quad u_{61} = 0$, $\quad u_{62} = -0.0602156728$,

$u_{63} = 0.0599893868$, $\quad u_{64} = -0.0167793321$.

The runtimes of the proposed algorithm for $\alpha(x, t) = 1 - 0.04 \cos^2(t) \sin^2(x)$ and $\alpha(x, t) = 1 - 0.8 \cos^2(t) \sin^2(x)$ are about 321 and 318 s, respectively.

This problem is also solved by the proposed method with $m_1 = 6$ and $m_2 = 3$. The absolute errors for $\alpha(x, t) = 1 - 0.04 \cos^2(t) \sin^2(x)$ (left) and $\alpha(x, t) = 1 - 0.8 \cos^2(t) \sin^2(x)$ (right) are shown in Fig. 3. From Theorems (2.2) and (2.3) we conclude that the new method is suitable for this problem and that by increasing the number of the GP, the accuracy of the result is improved. Observing the results presented in Table 2 and Figs. 2 and 3, we conclude that the proposed method provides approximate solutions with high accuracy.

Example 3 Consider the NV-TFCDE (1.1). The nonlinear operator H and the function $f(x, t)$ are:

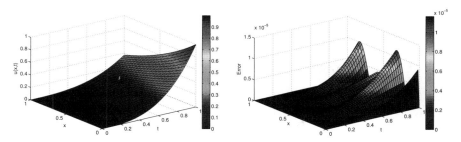

Fig. 2 The approximate solution (left) and absolute error (right) with $m_1 = 5$, $m_2 = 3$ and $\alpha(x, t) = 1 - 0.04 \cos^2(t) \sin^2(x)$ for Example 2

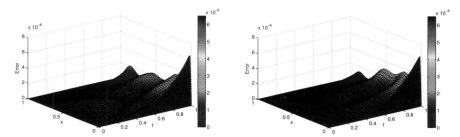

Fig. 3 The absolute errors for $\alpha(x, t) = 1 - 0.04 \cos^2(t) \sin^2(x)$ (left) and $\alpha(x, t) = 1 - 0.8 \cos^2(t) \sin^2(x)$ (right) with $(m_1 = 6, m_2 = 3)$ for Example 2

$$H(u(x, t)) = -u^3(x, t), \qquad f(x, t) = \frac{\Gamma\left(\frac{9}{2}\right) t^{\frac{7}{2} - \alpha(x, t)}}{\Gamma\left(\frac{9}{2} - \alpha(x, t)\right)} x^{\frac{5}{2}} - \frac{15}{4} t^{\frac{7}{2}} x^{\frac{1}{2}} + \frac{5\eta}{2} t^{\frac{7}{2}} x^{\frac{3}{2}} + \left(t^{\frac{7}{2}} x^{\frac{5}{2}}\right)^3.$$

The exact solution of this problem is $u(x, t) = t^{\frac{7}{2}} x^{\frac{5}{2}}$. The initial and boundary conditions can be computed using the exact solution. This problem is solved by the proposed method with $m_1 = 2$, $m_2 = 3$, $\eta = 2$ for the VO $\alpha(x, t) = 0.5 \pm 0.2 e^{-xt}$. The absolute errors $|e(x_i, t_i)|$ of the approximate solution in different points $(x, t) \in \Omega$ are shown in Table 3. The approximate solution and absolute error for $\alpha(x, t) = 0.5 + 0.2 e^{-xt}$ are represented in Fig. 4. The CP and FC for this problem with $\alpha(x, t) = 0.5 + 0.2 e^{-xt}$ are obtained as follows:

$s_1 = 1.4779857238, \qquad s_2 = 1.50, \qquad s_3 = 0.50, \qquad k_2 = 0.50,$

$u_{11} = 0, \qquad u_{12} = 0, \qquad u_{13} = 0, \qquad u_{14} = 0, \qquad u_{21} = 0, \qquad u_{22} = 0, \qquad u_{23} = -0.0170045691,$

$u_{24} = 0.0170045691, \qquad u_{31} = 0, \qquad u_{32} = 0, \qquad u_{33} = 0.4406251750, \qquad u_{34} = 0.5593748250.$

The runtimes of the proposed algorithm for $\alpha(x, t) = 0.5 + 0.2 e^{-xt}$ and $\alpha(x, t) = 0.5 - 0.2 e^{-xt}$ are about 226 and 219 s, respectively. Table 3 and Fig. 4 show that the proposed method provides approximate solutions with good accuracy.

Table 3 The absolute errors $|e(x_i, t_i)|$ of the approximate solutions in different points for Example 3

(x, t)	$\alpha(x, t) = 0.5 + 0.2e^{-xt}$	$\alpha(x, t) = 0.5 - 0.2e^{-xt}$
$(0.1, 0.1)$	5.5760E-20	2.4015E-20
$(0.2, 0.2)$	3.4452E-19	1.7165E-19
$(0.3, 0.3)$	6.9209E-19	4.2973E-19
$(0.4, 0.4)$	6.9607E-19	6.5775E-19
$(0.5, 0.5)$	1.3461E-19	7.0437E-19
$(0.6, 0.6)$	7.6690E-19	5.1468E-19
$(0.7, 0.7)$	1.3574E-18	1.8132E-19
$(0.8, 0.8)$	9.6221E-19	8.8979E-20
$(0.9, 0.9)$	2.6766E-19	1.1446E-19

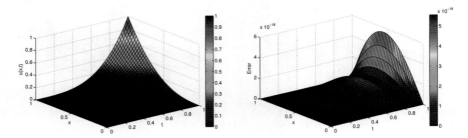

Fig. 4 The approximate solution (left) and absolute error (right) with $m_1 = 2$, $m_2 = 3$ and $\alpha(x, t) = 0.5 + 0.2e^{-xt}$ for Example 3

Example 4 Consider the NV-TFCDE (1.1) subject to the homogeneous initial and boundary conditions with the nonlinear operator H and the function $f(x, t)$:

$$H(u(x, t)) = u^4(x, t), \; f(x, t) = (x^2 - x^3)(4t^3 + t^4)e^t + (-2 + 8x - 3x^2)t^4e^t - ((x^2 - x^3)t^4e^t)^4.$$

The exact solution of this problem for $\alpha(x, t) = 1$ is $u(x, t) = (x^2 - x^3)t^4e^t$. This problem is also solved by the proposed method with $(m_1 = 3, m_2 = 4)$, $(m_1 = 4, m_2 = 6)$ and $\eta = 1$. The absolute errors for $(m_1 = 3, m_2 = 4)$ (left) and $(m_1 = 4, m_2 = 6)$ (right) with $\alpha(x, t) = 1$ are shown in Fig. 5. The approximate solutions for $\alpha(x, t) = 1$, $\alpha(x, t) = \frac{1}{2}(2 - \sin(x + t)/50)$ and $\alpha(x, t) = \frac{1 - (xt)^2}{50}$ for some different times and spaces are shown in Figs. 6 and 7, respectively. The CP and FC for this problem with $m_1 = 3$, $m_2 = 4$ and $\alpha(x, t) = 1$ are obtained as follows:

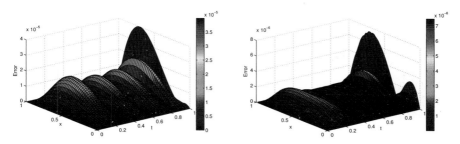

Fig. 5 The absolute errors with $(m_1 = 3, m_2 = 4)$ (left) and $(m_1 = 4, m_2 = 6)$ (right) for Example 4

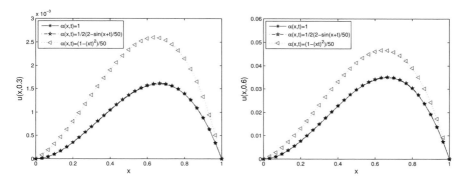

Fig. 6 The approximate solutions at $t = 0.3$ (left side) and $t = 0.6$ (right side) with $m_1 = 3, m_2 = 4$ for Example 4

$s_1 = 2.3003269777, \quad s_2 = 1.9875539508, \quad s_3 = 1.0906599116, \quad s_4 = 2.0587341446,$

$k_2 = 0.0000736908, \quad k_3 = -0.0000954315, \quad u_{11} = 0, \quad u_{12} = 0.0000016845, \quad u_{13} = -0.0000221026,$

$u_{14} = 0.0000209519, \quad u_{15} = 0, \quad u_{21} = 0, \quad u_{22} = 0.0046034305, \quad u_{23} = -0.0481579165,$

$u_{24} = 0.0440121319, \quad u_{25} = -0.0003861173, \quad u_{31} = 0, \quad u_{32} = -0.1240189371, \quad u_{33} = 0.6124468201,$

$u_{34} = 0.9992322750, \quad u_{35} = 1.2306308904, \quad u_{41} = 0, \quad u_{42} = 0.1194124856, \quad u_{43} = -0.5642494760,$

$u_{44} = -1.0432817712, \quad u_{45} = -1.2302438154.$

The runtime of the proposed algorithm for this example is about 138 seconds for $m_1 = 3, m_2 = 4$ with $\alpha(x, t) = 1$.

4 Conclusions

In this chapter, we presented an efficient and accurate optimization method based on the GP together with the operational matrices of the integer order and the V-FD for the NV-TFCDE. The method reduces the problem to a nonlinear system of algebraic

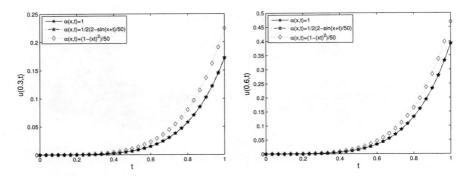

Fig. 7 The approximate solutions at $x = 0.3$ (left side) and $x = 0.6$ (right side) with $m_1 = 3$, $m_2 = 4$ for Example 4

equations for choosing the FC and CP optimally. Several numerical examples were given to test the applicability and efficiency of the new algorithm.

References

1. Pedro HTC, Kobayashi MH, Pereira JMC, Coimbra CFM (2008) Variable order modeling of diffusive-convective effects on the oscillatory flow past a sphere. J Vib Control 14:1569–1672
2. Ramirez LES, Coimbra C (2011) On the variable order dynamics of the nonlinear wake caused by a sedimenting particle. Physica D 240:1111–1118
3. Coimbra CFM (2003) Mechanics with variable-order differential operators. Ann Phys 12:692–703
4. Soon CM, Coimbra CFM, Kobayashi M (2005) The variable viscoelasticity oscillator. Ann Phys 14(6):378–389
5. Ramirez LES, Coimbra C, Kobayashi M (2007) Variable order constitutive relation for viscoelasticity. Ann Phys 16:543–552
6. Sun HG, Chen YQ, Chen W (2011) Random-order fractional differential equation models. Sign Process 91:525–530
7. Zahra WK, Hikal MM (2015) Non standard finite difference method for solving variable order fractional optimal control problems. J Vib Control:1–11
8. Shen S, Liu F, Chen J, Turner I, Anh V (2012) Numerical techniques for the variable order time fractional diffusion equation. Appl Math Comput 218:10861–10870
9. Yang XJ, Tenreiro Machado JA (2016) A new fractional operator of variable order: application in the description of anomalous diffusion. arXiv preprint arXiv:1611.09200
10. Bhrawy AH, Zaky MA (2015) Numerical simulation for two-dimensional variable-order fractional nonlinear cable equation. Nonlinear Dynam 80:101–116
11. Zayernouri M, Karniadakis GE (2015) Fractional spectral collocation methods for linear and nonlinear variable order FPDEs. J Comput Phys 239:312–338
12. Bhrawy AH, Zaky MA (2017) An improved collocation method for multi-dimensional space-time variable-order fractional Schrödinger equations. Appl Numer Math 11:197–218
13. Bhrawy AH, Zaky MA (2016) Numerical algorithm for the variable-order Caputo fractional functional differential equation. Nonlinear Dynam 85:1815–1823
14. Zhang H, Liu F, Phanikumar MS, Meerschaert MM (2013) A novel numerical method for the time variable fractional order mobile-immobile advection-dispersion model. Comput Math Appl 66:693–701

15. Zhang H, Liu F, Zhuang P, Turner I, Anh V (2014) Numerical analysis of a new space-time variable fractional order advection-dispersion equation. Appl Math Comput 242:541–550
16. Chen S, Liu F, Burrage K (2014) Numerical simulation of a new two-dimensional variable-order fractional percolation equation in non-homogeneous porous media. Comput Math Appl 68(12):2133–2141
17. Chen Y, Liu L, Li B, Sun Y (2014) Numerical solution for the variable order linear cable equation with Bernstein polynomials. Appl Math Comput 238:329–341
18. Abdelkawy MA, Zaky MA, Bhrawy AH, Baleanu D (2015) Numerical simulation of time variable fractional order mobile-immobile advection-dispersion model. Romanian Rep Phys 67(3):773–791
19. Chen YM, Wei YQ, Liu DY, Yu H (2015) Numerical solution for a class of nonlinear variable order fractional differential equations with Legendre wavelets. Appl Math Lett 46:83–88
20. Chen YM, Wei YQ, Liu DY, Boutat D, Chen XK (2016) Variable-order fractional numerical differentiation for noisy signals by wavelet denoising. J Comput Phys 311:338–347
21. Zhao X, Sun ZZ, Karniadakis GE (2015) Second-order approximations for variable order fractional derivatives: algorithms and applications. J Comput Phys 293:184–200
22. Li XY, Wu B (2015) A numerical technique for variable fractional functional boundary value problems. Appl Math Lett 43:108–113
23. Jia YT, Xu MQ, Lin YZ (2017) A numerical solution for variable order fractional functional differential equation. Appl Math Lett 64:125–130
24. Chen CM (2013) Numerical methods for solving a two-dimensional variable-order modified diffusion equation. Appl Math Comput 225:62–78
25. Dabiri A, Parsa Moghaddam B, Tenreiro Machado JA (2018) Optimal variable-order fractional PID controllers for dynamical systems. J Comput Appl Math 339:40–48
26. Nagy AM, Sweilam NH (2018) Numerical simulations for a variable order fractional cable equation. Acta Math Sci 38(2):580–590
27. Malesza W, Macias M, Sierociuk D (2019) Analytical solution of fractional variable order differential equations. J Comput Appl Math 348:214–236
28. Hassani H, Avazzadeh Z, Tenreiro Machado JA (2019) Solving two dimensional variable-order fractional optimal control problems with transcendental Bernstein series. J Comput Nonlinear Dyn 19:061001–11
29. Hassani H, Naraghirad E (2019) A new computational method based on optimization scheme for solving variable-order time fractional Burgers' equation. Math Comput Simul 162:1–17
30. Hassani H, Tenreiro Machado JA, Avazzadeh Z (2019) An effective numerical method for solving nonlinear variable-order fractional functional boundary value problems through optimization technique. Nonlinear Dynam 97(4):2041–2054
31. Mohammadi F, Hassani H (2019) Numerical solution of two-dimensional variable-order fractional optimal control problem by generalized polynomial basis. J Optim Theory Appl 180(2):536–555
32. Meng R, Yin D, Drapaca CS (2019) Variable-order fractional description of compression deformation of amorphous glassy polymers. Comput Mech 64(1):163–171
33. Malesz W, Macias M, Sierociuk D (2019) Analytical solution of fractional variable order differential equations. J Comput Appl Math 348:214–236
34. Dahaghin MSh, Hassani H (2017) An optimization method based on the generalized polynomials for nonlinear variable-order time fractional diffusion-wave equation. Nonlinear Dynam 88(3):1587–1598

Inverse Problems for Some Systems
of Parabolic Equations

C. Connell McCluskey and Vitali Vougalter

Abstract We study the system $\vec{u}_t - A\vec{u}_{xx} = \vec{h}(t)$, where $0 \leq x \leq \pi$, $t \geq 0$, assuming that $\vec{u}(0, t) = \vec{v}(t)$, $\vec{u}(\pi, t) = \vec{0}$, and $\vec{u}(x, 0) = \vec{g}(x)$. The coupling matrix A is a real, diagonalizable matrix for which all of the eigenvalues are positive reals. The question is: *What extra data determine the three unknown vector functions* $\{\vec{h}, \vec{v}, \vec{g}\}$ *uniquely?* This problem is solved and an analytical method for the recovery of the above three vector functions is presented.

Keywords Parabolic systems · Inverse problems · Inverse source problems

AMS Subject Classification: 35K20 · 35R30

1 Introduction

Consider the system

$$\vec{u}_t - A\vec{u}_{xx} = \vec{h}(t) \quad \text{for } (x, t) \in [0, \pi] \times [0, \infty),$$
$$\text{with } \vec{u}(0, t) = \vec{v}(t), \quad \vec{u}(\pi, t) = \vec{0}, \text{ and } \vec{u}(x, 0) = \vec{g}(x), \tag{1.1}$$

where $\vec{h}, \vec{v} : \mathbb{R}_{\geq 0} \to \mathbb{R}^N$ and $\vec{g} : [0, \pi] \to \mathbb{R}^N$ for some $N \geq 2$ are unknown, with

$$\vec{g}(x) = (g_1(x), g_2(x), ..., g_N(x))^T.$$

The solution of system (1.1) is a real vector function given by

C. Connell McCluskey (✉)
Department of Mathematics, Wilfrid Laurier University, Waterloo, ON N2L 3C5, Canada

V. Vougalter
Department of Mathematics, University of Toronto, Toronto, ON M5S 2E4, Canada
e-mail: vitali@math.toronto.edu

$$\vec{u}(x, t) = (u_1(x, t), u_2(x, t), ..., u_N(x, t))^T.$$

The regularity of $\vec{u}$ is related to the smoothness of $\{\vec{h}, \vec{v}, \vec{g}\}$. Similar to the work in [4], which was devoted to the studies of the single parabolic equation of this kind, in the present work we are not focused on the well-posedness of (1.1). We are interested in the following inverse problem:

> What information about the solution $\vec{u}$ is sufficient
> to uniquely determine the vector functions $\{\vec{h}, \vec{v}, \vec{g}\}$?

Inverse problems for the scalar heat equation have been studied extensively (see [1, 2, 4] and the references therein). An inverse source problem for the multidimensional heat equation in which the source was assumed to be a finite sum of point sources was considered in [3]. The inverse problem there was to find the location and the intensity of these point sources from the experimental data. The existence of stationary solutions of certain systems of parabolic equations was studied actively in recent years, see for instance [5, 6] and the references therein.

We will use $\langle \, , \, \rangle$ to denote the standard inner product on $L^2[0, \pi]$. That is,

$$\langle G, F \rangle = \int_0^\pi G(x)F(x)dx. \tag{1.2}$$

Clearly, (1.2) induces the following norm on $L^2[0, \pi]$:

$$\|F\| = \sqrt{\int_0^\pi F^2(x)dx}.$$

We extend the inner product notation to the situation where the first argument is a vector function, for which each component is an element of $L^2[0, \pi]$. In this case the result is obtained by computing the inner product of each component with the second argument. For example,

$$\langle \vec{g}, F \rangle = \left(\langle g_1, F \rangle, \ldots, \langle g_N, F \rangle \right)^T$$
$$= \left(\int_0^\pi g_1(x)F(x)dx, \ldots, \int_0^\pi g_N(x)F(x)dx \right)^T \tag{1.3}$$
$$= \int_0^\pi \vec{g}(x)F(x)dx.$$

Similarly,

$$\langle \vec{u}(\cdot, t), F \rangle = \int_0^\pi \vec{u}(x, t)F(x)dx$$
$$= \left(\int_0^\pi u_1(x, t)F(x)dx, \ldots, \int_0^\pi u_N(x, t)F(x)dx \right)^T,$$

giving a vector valued function of t.

Let $f_m(x) = \sqrt{\dfrac{2}{\pi}} \sin(mx)$ for $m \in \mathbb{N} = \{1, 2, \dots\}$. Then

$$f_m(0) = f_m(\pi) = 0, \quad \|f_m\| = 1 \text{ and } -\frac{d^2 f_m}{dx^2}(x) = m^2 f_m(x) \text{ for } 0 \le x \le \pi,$$

so that $\{f_m(x)\}_{m=1}^{\infty}$ is the orthonormal set of the eigenfunctions of the one dimensional negative Dirichlet Laplacian on the interval $[0, \pi]$.

Let $y \in (0, \pi)$ such that $\dfrac{y}{\pi}$ is irrational. (This happens, for example, if y is rational.) Then it can be shown that

$$f_m(y) \neq 0 \tag{1.4}$$

for all $m \in \mathbb{N}$.

Let

$$\vec{u}_m(t) = \langle \vec{u}(\cdot, t), f_m \rangle$$

for $m \in \mathbb{N}$. Our main statement is as follows.

Theorem 1 *Suppose $N \ge 2$ and A is a constant real $N \times N$ diagonalizable matrix for which all of the eigenvalues are positive reals. Then knowing the functions*

$$\{\vec{u}_1(t), \vec{u}_3(t), \vec{u}(y, t)\}, \tag{1.5}$$

for all $t \ge 0$, is sufficient to uniquely determine the triple $\left\{\vec{h}, \vec{v}, \vec{g}\right\}$.

This theorem is a generalization of Theorem 1 of [4], which establishes the corresponding result for a single heat equation (i.e. for $N = 1$). Let us proceed to the proof of our main result.

2 Proof

Proof of Theorem 1. From our assumptions, it follows that there exists an invertible real matrix P such that

$$PAP^{-1} = D = \text{diag}(d_1, d_2, \dots, d_N),$$

where $d_1, d_2, \dots, d_N > 0$ are the eigenvalues of A and, hence,

$$PA = DP. \tag{2.1}$$

By means of (2.1), multiplying the partial differential equation in (1.1) on the left by P gives

$$P\vec{u}_t - DP\vec{u}_{xx} = P\vec{h}(t). \tag{2.2}$$

Let us introduce new vector functions:

$$\tilde{u}(x, t) := P\vec{u}(x, t) \text{ and } \tilde{h}(t) := P\vec{h}(t).$$

This allows us to write (2.2) (which is simply the PDE portion of the main system (1.1)) in terms of $\tilde{u}$ and $\tilde{h}$. Before doing so, we define

$$\tilde{v}(t) := P\vec{v}(t) \text{ and } \tilde{g}(x) := P\vec{g}(x).$$

Now we can write the system in terms of the new variables:

$$\frac{\partial \tilde{u}}{\partial t} - D\frac{\partial^2 \tilde{u}}{\partial x^2} = \tilde{h}(t), \tag{2.3}$$

$$\text{with} \quad \tilde{u}(0, t) = \tilde{v}(t), \qquad \tilde{u}(\pi, t) = \vec{0} \text{ and } \tilde{u}(x, 0) = \tilde{g}(x).$$

The reason that we have done this is that (2.3) consists of N fully decoupled scalar equations, allowing for solutions to be more easily obtained.

For $m \in \mathbb{N} = \{1, 2, \dots\}$ let

$$\tilde{u}_m(t) = \langle \tilde{u}(\cdot, t), f_m \rangle \text{ and } \tilde{g}_m = \langle \tilde{g}, f_m \rangle \in \mathbb{R}^N,$$

where the inner product is defined in (1.3). It follows that

$$\tilde{g}(x) = \sum_{m=1}^{\infty} \tilde{g}_m f_m(x). \tag{2.4}$$

We look for the solution to (2.3) in the form

$$\tilde{u}(x, t) = \sum_{m=1}^{\infty} \tilde{u}_m(t) f_m(x) = \sum_{m=1}^{\infty} \left\langle \tilde{u}(\cdot, t), f_m \right\rangle f_m(x). \tag{2.5}$$

It is a standard result that such a solution exists. Taking the inner product of f_m with each side of the system of partial differential equations in (2.3) yields

$$\left\langle \frac{\partial \tilde{u}}{\partial t} - D\frac{\partial^2 \tilde{u}}{\partial x^2}, f_m \right\rangle = \tilde{h}(t)\langle \mathbf{1}, f_m \rangle, \tag{2.6}$$

where $\mathbf{1}(x) = 1$ for all $x \in [0, \pi]$. Letting

$$c_m = \langle \mathbf{1}, f_m \rangle = \int_0^{\pi} f_m(x)dx = \begin{cases} \sqrt{\frac{2}{\pi}}\frac{2}{m} & \text{if } m \text{ is odd} \\ 0 & \text{if } m \text{ is even,} \end{cases} \tag{2.7}$$

we rewrite (2.6) as

$$\int_0^\pi \frac{\partial \tilde{u}}{\partial t}(x, t) f_m(x)dx - \int_0^\pi D\frac{\partial^2 \tilde{u}}{\partial x^2}(x, t) f_m(x)dx = \tilde{h}(t)c_m.$$

Assuming the sufficient regularity of $\tilde{u}$, the first integral gives $\dfrac{d\tilde{u}_m}{dt}$. Using integration by parts twice on the second integral, we arrive at

$$\frac{d\tilde{u}_m(t)}{dt} + Dm^2\tilde{u}_m(t) = Df_m'(0)\tilde{v}(t) + c_m\tilde{h}(t), \qquad (2.8)$$

for $m \in \mathbb{N}$. Equation (2.8) decouples into N scalar linear equations of the form $y' + Ky = a(t)$, which can be easily solved. The initial condition for (2.8) is

$$\tilde{u}_m(0) = \langle \tilde{u}(\cdot, 0), f_m \rangle = \langle \tilde{g}, f_m \rangle = \tilde{g}_m. \qquad (2.9)$$

Recall that for a diagonal matrix, such as $D = \text{diag}\,(d_1, \ldots, d_N)$, and a scalar $m^2 t$, exponentiation is termwise, so that

$$e^{-Dm^2 t} = \text{diag}\left(e^{-d_1 m^2 t}, \ldots, e^{-d_N m^2 t}\right).$$

From (2.8) we calculate that

$$\tilde{u}_m(t) = e^{-Dm^2 t}\tilde{g}_m + \int_0^t e^{-Dm^2(t-s)}\left[Df_m'(0)\tilde{v}(s) + c_m\tilde{h}(s)\right]ds. \qquad (2.10)$$

We now suppose that
$$\{\vec{u}_1(t), \vec{u}_3(t), \vec{u}(y, t)\}$$

(referred to as the data) are known, and we set about constructing the unknowns $\left\{\vec{h}, \vec{v}, \vec{g}\right\}$. Let

$$\tilde{F}_1(t) := \tilde{u}_1(t) - e^{-Dt}\tilde{g}_1 \text{ and } \tilde{F}_3(t) := \tilde{u}_3(t) - e^{-9Dt}\tilde{g}_3. \qquad (2.11)$$

Then Eq. (2.10), for $m = 1$ and $m = 3$, gives

$$\tilde{F}_1(t) = \int_0^t e^{-D(t-s)}\left[Df_1'(0)\tilde{v}(s) + c_1\tilde{h}(s)\right]ds$$

$$\text{and } \tilde{F}_3(t) = \int_0^t e^{-9D(t-s)}\left[Df_3'(0)\tilde{v}(s) + c_3\tilde{h}(s)\right]ds \qquad (2.12)$$

respectively. By differentiating the formulas in (2.12) and rearranging, we obtain

$$Df_1'(0)\tilde{v}(t) + c_1\tilde{h}(t) = e^{-Dt}\frac{d}{dt}\left[e^{Dt}\tilde{F}_1(t)\right],$$

$$\text{and } Df_3'(0)\tilde{v}(t) + c_3\tilde{h}(t) = e^{-9Dt}\frac{d}{dt}\left[e^{9Dt}\tilde{F}_3(t)\right]. \tag{2.13}$$

We treat (2.13) as a $2N$-dimensional linear system with unknowns $\tilde{v}(t)$ and $\tilde{h}(t)$. Its $2N \times 2N$ coefficient matrix M has the block form

$$M = \begin{pmatrix} Df_1'(0) & c_1I \\ Df_3'(0) & c_3I \end{pmatrix},$$

where I is the $N \times N$ identity matrix and $f_m'(0) = m\sqrt{\frac{2}{\pi}}$. Since each of the four blocks in this representation of M are diagonal, an easy computation gives $\det M = \left(\frac{32}{3\pi}\right)^N \det D \neq 0$. It now follows that system (2.13) admits a unique solution, meaning that $\tilde{v}$ and $\tilde{h}$ are uniquely determined by the right-hand sides of (2.13).

Note that for $m = 1, 3$, we have

$$\tilde{u}_m(t) = \langle \tilde{u}(\cdot, t), f_m \rangle = \langle P\vec{u}(\cdot, t), f_m \rangle = \int_0^\pi P\vec{u}(x, t)f_m(x)dx$$

$$= P\int_0^\pi \vec{u}(x, t)f_m(x)dx = P\vec{u}_m(t).$$

Also, from the definition of $\tilde{g}_m$ in (2.9), we have

$$\tilde{g}_m = \tilde{u}_m(0) = P\vec{u}_m(0).$$

This means that $\tilde{F}_1(t)$ and $\tilde{F}_3(t)$, as defined in (2.11), can be calculated from the data in (1.5). This, in turn, means that $\tilde{v}(t)$ and $\tilde{h}(t)$ can be computed from the data. We then calculate

$$\vec{v}(t) = P^{-1}\tilde{v}(t) \text{ and } \vec{h}(t) = P^{-1}\tilde{h}(t).$$

Now we work towards determining $\vec{g}$ in terms of the data. By combining (2.5) and (2.10) at $x = y$, we have

$$\tilde{u}(y, t) = \sum_{m=1}^\infty e^{-Dm^2t}\tilde{g}_m f_m(y) + \tilde{w}(y, t), \tag{2.14}$$

where

$$\tilde{w}(y, t) = \sum_{m=1}^\infty f_m(y)\int_0^t e^{-Dm^2(t-s)}[Df_m'(0)\tilde{v}(s) + c_m\tilde{h}(s)]ds$$

an expression written in terms of known functions (since $\tilde{v}$ and $\tilde{h}$ have already been calculated from the data). Also, noting that $\tilde{u}(y, t) = P\vec{u}(y, t)$, it is clear that the left side of (2.14) is determined by the data (1.5). Thus, the only unknowns in (2.14) are $\tilde{g}_m$ for $m \in \mathbb{N}$.

Let $\tilde{q}(y, t) = \tilde{u}(y, t) - \tilde{w}(y, t)$. Then $q(y, t)$ is known as well and, from (2.14), satisfies

$$\tilde{q}(y, t) = \sum_{m=1}^{\infty} e^{-Dm^2 t} \tilde{g}_m f_m(y)$$

$$= e^{-Dt} \tilde{g}_1 f_1(y) + e^{-4Dt} \tilde{g}_2 f_2(y) + e^{-9Dt} \tilde{g}_3 f_3(y) + \cdots .$$

(2.15)

We now perform a sequence of limits. The first limit is simply $e^{Dt} \tilde{q}(y, t)$ as t approaches ∞, which equals $\tilde{g}_1 f_1(y)$. That is,

$$\tilde{g}_1 f_1(y) = \lim_{t \to \infty} \left[e^{Dt} \tilde{q}(y, t) \right].$$

For the second limit, we subtract the first term of the series in (2.15) to the other side and multiply by e^{4Dt}, so that the limit gives $\tilde{g}_2 f_2(y)$. That is,

$$\tilde{g}_2 f_2(y) = \lim_{t \to \infty} e^{4Dt} \left[q(y, t) - e^{-Dt} \tilde{g}_1 f_1(y) \right].$$

Continuing in this fashion, we are able to calculate $\tilde{g}_m f_m(y)$ for each $m \in \mathbb{N}$. By Eq. (1.4), each $f_m(y)$ is non-zero and so $\tilde{g}_m$ has now been determined for each $m \in \mathbb{N}$. Then, by (2.4) the vector function $\tilde{g}$ is determined. Finally, $\vec{g}(x) = P^{-1} \tilde{g}(x)$.

Thus, the triple $\left\{ \vec{h}(t), \vec{v}(t), \vec{g}(x) \right\}$ is uniquely determined and can in fact be calculated from the given data $\{\vec{u}_1(t), \vec{u}_3(t), \vec{u}(y, t)\}$, $t \geq 0$. ∎

Remark 1 We note that the initial data could be $\{\vec{u}_i(t), \vec{u}_j(t), \vec{u}(y, t)\}$, as long as the resulting matrix M is non-singular, which is the case as long as the 2×2 matrix

$$M^* = \begin{pmatrix} f_i'(0) & c_i \\ f_j'(0) & c_j \end{pmatrix}$$

is non-singular. Noting that $f_m'(0) = m\sqrt{\dfrac{2}{\pi}}$ and that c_m is given in (2.7), it follows that M^* is non-singular as long as $i \neq j$ and at least one of i and j is odd. In such a case, it would still be possible to calculate $\left\{ \vec{h}(t), \vec{v}(t), \vec{g}(x) \right\}$ from the data.

A reasonable way to illustrate the method presented above more explicitly would be to consider instead of system (1.1) a scalar equation of this kind. This has been done in [4].

Acknowledgements Valuable discussions with A.G. Ramm are gratefully acknowledged. The work was partially supported by the NSERC Discovery grant.

References

1. Ramm AG (2001) An inverse problem for the heat equation. J Math Anal Appl 264(2):691–697
2. Ramm AG (2005a) Inverse problems. Mathematical and analytical techniques with applications to engineering. Springer, New York, p 442
3. Ramm AG (2005b) Inverse problems for parabolic equations. Aust J Math Anal Appl 2(2):5
4. Ramm AG (2007) Inverse problems for parabolic equations II. Commun Nonlinear Sci Numer Simul 12(6):865–868
5. Vougalter V, Volpert V (2017) Existence of stationary solutions for some systems of integro-differential equations with superdiffusion. Rocky Mountain J Math 47(3):955–970
6. Vougalter V, Volpert V (2018) On the existence in the sense of sequences of stationary solutions for some systems of non-Fredholm integro-differential equations. Mediterr J Math 15(5):19

Solvability in the Sense of Sequences for Some Non-Fredholm Operators with Drift

Vitaly Volpert and Vitali Vougalter

Abstract We study solvability of certain linear nonhomogeneous elliptic equations and prove that under reasonable technical conditions the convergence in L^2 of their right sides yields the existence and the convergence in H^2 of the solutions. The problems involve second order differential operators with or without Fredholm property, on the whole real line or on a finite interval with periodic boundary conditions. We show that the drift term involved in these equations provides the regularization for the solutions of our problems.

Keywords Solvability conditions · Non-Fredholm operators · Sobolev spaces · Drift term

Classifications 35A01 · 35P10 · 47F05

1 Introduction

Let us consider the problem

$$- \Delta u + V(x)u - au = f, \tag{1.1}$$

where $u \in E = H^2(\mathbb{R}^d)$ and $f \in F = L^2(\mathbb{R}^d)$, $d \in \mathbb{N}$, a is a constant and $V(x)$ is a function decaying to 0 at infinity. If $a \geq 0$, then the essential spectrum of the operator $A : E \to F$, which corresponds to the left-hand side of Eq. (1.1) contains the origin. Consequently, such operator fails to satisfy the Fredholm property. Its image is not closed, for $d > 1$ the dimensions of its kernel and the codimension of its image

---

V. Volpert (✉)
Institute Camille Jordan, UMR 5208 CNRS, University Lyon 1, 69622 Villeurbanne, France
e-mail: volpert@math.univ-lyon1.fr

V. Vougalter
Department of Mathematics, University Toronto, Toronto, ON M5S 2E4, Canada
e-mail: vitali@math.toronto.edu

are not finite. In this article we will study some properties of such operators. Note that elliptic equations involving non-Fredholm operators were treated extensively in recent years (see [3, 11–14, 16–19]) along with their potential applications to the theory of reaction-diffusion equations (see [7, 8]). In the particular case where $a = 0$ the operator A satisfies the Fredholm property in some properly chosen weighted spaces [1–3, 5, 6]. However, the case with $a \neq 0$ is essentially different and the method developed in these articles cannot be applied.

One of the important issues about equations with non-Fredholm operators concerns their solvability. We will study it in the following setting. Let f_n be a sequence of functions in the image of the operator A, such that $f_n \to f$ in $L^2(\mathbb{R}^d)$ as $n \to \infty$. Denote by u_n a sequence of functions from $H^2(\mathbb{R}^d)$ such that

$$Au_n = f_n, \ n \in \mathbb{N}.$$

Since the operator A fails to satisfy the Fredholm property, the sequence u_n may not be convergent. Let us call a sequence u_n such that $Au_n \to f$ a solution in the sense of sequences of equation $Au = f$ (see [10]). If this sequence converges to a function u_0 in the norm of the space E, then u_0 is a solution of this equation. Solution in the sense of sequences is equivalent in this sense to the usual solution. However, in the case of non-Fredholm operators this convergence may not hold or it can occur in some weaker sense. In such case, solution in the sense of sequences may not imply the existence of the usual solution. In this work we will find sufficient conditions of equivalence of solutions in the sense of sequences and the usual solutions. In the other words, the conditions on sequences f_n under which the corresponding sequences u_n are strongly convergent.

In the first part of the work, we consider the equation with the drift term

$$-\frac{d^2u}{dx^2} - b\frac{du}{dx} - au = f(x), \quad x \in \mathbb{R}, \tag{1.2}$$

where $a \geq 0$ and $b \in \mathbb{R}$, $b \neq 0$ are constants and the right side is square integrable. The equation with drift in the context of the Darcy's law describing the fluid motion in the porous medium was treated in [17]. The drift term arises when studying the emergence and propagation of patterns arising in the theory of speciation (see [20]). Nonlinear propagation phenomena for the reaction-diffusion type equations involving the drift term was studied in [4]. The operator involved in the left side of (1.2)

$$L_{a,b} := -\frac{d^2}{dx^2} - b\frac{d}{dx} - a : \quad H^2(\mathbb{R}) \to L^2(\mathbb{R}) \tag{1.3}$$

is non-selfadjoint. By means of the standard Fourier transform

$$\widehat{f}(p) := \frac{1}{\sqrt{2\pi}} \int_{-\infty}^{\infty} f(x)e^{-ipx}dx, \quad p \in \mathbb{R} \tag{1.4}$$

it can be easily verified that the essential spectrum of the operator $L_{a,\,b}$ is given by

$$\lambda_{a,\,b}(p) := p^2 - a - ibp, \quad p \in \mathbb{R}.$$

Evidently, when $a > 0$ the operator $L_{a,\,b}$ is Fredholm, because its essential spectrum does not contain the origin. But when $a = 0$ the operator $L_{0,\,b}$ is non-Fredholm since its essential spectrum contains the origin.

Note that in the absence of the drift term we are dealing with the self-adjoint operator

$$-\frac{d^2}{dx^2} - a : \quad H^2(\mathbb{R}) \to L^2(\mathbb{R}), \quad a > 0,$$

which fails to satisfy the Fredholm property (see [19, 21]). Let us write down the corresponding sequence of iterated equations with $m \in \mathbb{N}$ as

$$-\frac{d^2 u_m}{dx^2} - b\frac{du_m}{dx} - au_m = f_m(x), \quad x \in \mathbb{R}, \tag{1.5}$$

where the right sides converge to the right side of (1.2) in $L^2(\mathbb{R})$ as $m \to \infty$. The inner product of two functions

$$(f(x), g(x))_{L^2(\mathbb{R})} := \int_{-\infty}^{\infty} f(x)\bar{g}(x)dx, \tag{1.6}$$

with a slight abuse of notations when these functions are not square integrable. Indeed, if $f(x) \in L^1(\mathbb{R})$ and $g(x) \in L^\infty(\mathbb{R})$, then clearly the integral considered above makes sense, like for example in the case of functions involved in the orthogonality relations (1.8) and (1.9) of Theorems 1 and 2 below. For our problem on the finite interval $I := [0, 2\pi]$ with periodic boundary conditions, we will use inner product analogously to (1.6), replacing the real line with I. In the article we will consider the space $H^2(\mathbb{R})$ equipped with the norm

$$\|u\|_{H^2(\mathbb{R})}^2 := \|u\|_{L^2(\mathbb{R})}^2 + \left\|\frac{d^2 u}{dx^2}\right\|_{L^2(\mathbb{R})}^2. \tag{1.7}$$

Our first main proposition is as follows.

Theorem 1 Let $f(x) : \mathbb{R} \to \mathbb{R}$ and $f(x) \in L^2(\mathbb{R})$.

(a) When $a > 0$ Eq. (1.2) admits a unique solution $u(x) \in H^2(\mathbb{R})$.
(b) When $a = 0$ let in addition $xf(x) \in L^1(\mathbb{R})$. Then problem (1.2) possesses a unique solution $u(x) \in H^2(\mathbb{R})$ if and only if the orthogonality condition

$$(f(x), 1)_{L^2(\mathbb{R})} = 0 \tag{1.8}$$

holds.

Note that the expression in the left side of (1.8) is well defined via the elementary argument analogous to the proof of Fact 1 of [14]. Then we turn our attention to proving the solvability in the sense of sequences for our equation on the real line.

Theorem 2 *Let $m \in \mathbb{N}$, $f_m(x) : \mathbb{R} \to \mathbb{R}$ and $f_m(x) \in L^2(\mathbb{R})$. Moreover, $f_m(x) \to f(x)$ in $L^2(\mathbb{R})$ as $m \to \infty$.*

(a) *When $a > 0$ Eqs. (1.2) and (1.5) have unique solutions $u(x) \in H^2(\mathbb{R})$ and $u_m(x) \in H^2(\mathbb{R})$ respectively, such that $u_m(x) \to u(x)$ in $H^2(\mathbb{R})$ as $m \to \infty$.*
(b) *When $a = 0$ let in addition $x f_m(x) \in L^1(\mathbb{R})$, such that $x f_m(x) \to x f(x)$ in $L^1(\mathbb{R})$ as $m \to \infty$. Furthermore,*

$$(f_m(x), 1)_{L^2(\mathbb{R})} = 0, \quad m \in \mathbb{N} \tag{1.9}$$

holds. Then problems (1.2) and (1.5) admit unique solutions $u(x) \in H^2(\mathbb{R})$ and $u_m(x) \in H^2(\mathbb{R})$ respectively, such that $u_m(x) \to u(x)$ in $H^2(\mathbb{R})$ as $m \to \infty$.

Note that in the parts (a) of Theorems 1 and 2 above the orthogonality conditions are not required, as distinct from the situation without a drift term discussed in the part (a) of Lemma 5 of [19] and in the part (a) of Theorem 1.1 of [21]. In the part b) of Theorems 1 and 2 of the present article only a single orthogonality condition is required, as distinct from the situation when $b = 0$ studied in the part b) of Lemma 5 of [19] and in the part (a) of Theorem 1.1 of [21], where there solvability was based on the two orthogonality relations. These facts show that the introduction of the drift term provides the regularization for the solutions of our problems.

In the second part of the work, we study the analogous equation on the finite interval with periodic boundary conditions, i.e. $I := [0, 2\pi]$, namely

$$-\frac{d^2 u}{dx^2} - b\frac{du}{dx} - au = f(x), \quad x \in I \tag{1.10}$$

where $a \geq 0$ and $b \in \mathbb{R}$, $b \neq 0$ are constants and the right side is bounded and periodic. Clearly,

$$\|f\|_{L^1(I)} \leq 2\pi \|f\|_{L^\infty(I)} < \infty, \quad \|f\|_{L^2(I)} \leq \sqrt{2\pi} \|f\|_{L^\infty(I)} < \infty, \tag{1.11}$$

such that $f(x) \in L^1(I) \cap L^2(I)$ as well. We will use the Fourier transform

$$f_n := \frac{1}{\sqrt{2\pi}} \int_0^{2\pi} f(x) e^{-inx} dx, \quad n \in \mathbb{Z}, \tag{1.12}$$

such that

$$f(x) = \sum_{n=-\infty}^{\infty} f_n \frac{e^{inx}}{\sqrt{2\pi}}.$$

Evidently, the non-selfadjoint operator involved in the left side of (1.10)

$$\mathcal{L}_{a,\,b} := -\frac{d^2}{dx^2} - b\frac{d}{dx} - a : \quad H^2(I) \to L^2(I) \tag{1.13}$$

is Fredholm. By applying (1.12), it can be easily verified that the spectrum of $\mathcal{L}_{a,\,b}$ is given by

$$\lambda_{a,\,b}(n) := n^2 - a - ibn, \quad n \in \mathbb{Z}$$

and the corresponding eigenfunctions are the Fourier harmonics $\dfrac{e^{inx}}{\sqrt{2\pi}}$, $n \in \mathbb{Z}$. Obviously, the eigenvalues of the operator $\mathcal{L}_{a,\,b}$ are simple, as distinct from the situation without the drift term, when the eigenvalues corresponding to $n \neq 0$ are doubly degenerate (see [15]). The appropriate functional space here $H^2(I)$ is

$$\{u(x) : I \to \mathbb{R} \mid u(x), u''(x) \in L^2(I), \quad u(0) = u(2\pi), \quad u'(0) = u'(2\pi)\}.$$

For the technical purposes, we will use the following auxiliary constrained subspace

$$H_0^2(I) = \{u(x) \in H^2(I) \mid (u(x), 1)_{L^2(I)} = 0\}, \tag{1.14}$$

which is a Hilbert spaces as well (see e.g. Chap. 2 of [9]). When $a > 0$, the kernel of the operator $\mathcal{L}_{a,\,b}$ is empty. For $a = 0$, we consider

$$\mathcal{L}_{0,\,b} : \quad H_0^2(I) \to L^2(I).$$

Such operator has the empty kernel as well. We write down the corresponding sequence of iterated equations with $m \in \mathbb{N}$ as

$$-\frac{d^2 u_m}{dx^2} - b\frac{du_m}{dx} - au_m = f_m(x), \quad x \in I, \tag{1.15}$$

where the right sides are bounded, periodic and converge to the right side of (1.10) in $L^\infty(I)$ as $m \to \infty$. Theorems 3 and 4 below are given to show the formal similarity of the results on the finite interval with periodic boundary conditions to the ones established for the whole real line case in Theorems 1 and 2.

Theorem 3 *Let $f(x) : I \to \mathbb{R}$, such that $f(0) = f(2\pi)$ and $f(x) \in L^\infty(I)$.*

(a) When $a > 0$ Eq. (1.10) has a unique solution $u(x) \in H^2(I)$.
(b) When $a = 0$ problem (1.10) admits a unique solution $u(x) \in H_0^2(I)$ if and only if the orthogonality condition

$$(f(x), 1)_{L^2(I)} = 0 \tag{1.16}$$

holds.

Our final main proposition deals with the solvability in the sense of sequences for our equation on the finite interval I.

Theorem 4 *Let $m \in \mathbb{N}$, $f_m(x) : I \to \mathbb{R}$, such that $f_m(0) = f_m(2\pi)$. Moreover, $f_m(x) \in L^\infty(I)$ and $f_m(x) \to f(x)$ in $L^\infty(I)$ as $m \to \infty$.*

(a) When $a > 0$ Eqs. (1.10) and (1.15) have unique solutions $u(x) \in H^2(I)$ and $u_m(x) \in H^2(I)$ respectively, such that $u_m(x) \to u(x)$ in $H^2(I)$ as $m \to \infty$.
(b) When $a = 0$ let

$$(f_m(x), 1)_{L^2(I)} = 0, \quad m \in \mathbb{N}. \tag{1.17}$$

Then (1.10) and (1.15) admit unique solutions $u(x) \in H_0^2(I)$ and $u_m(x) \in H_0^2(I)$ respectively, such that $u_m(x) \to u(x)$ in $H_0^2(I)$ as $m \to \infty$.

Note that in the parts (a) of Theorems 3 and 4 above the orthogonality relations are not required, as distinct from the situation when $a = n_0^2$, $n_0 \in \mathbb{N}$ considered in part II) of Theorem 2 of [15], see also part II) of Theorem 2.2 of [22], where the two orthogonality conditions were involved.

2   The Whole Real Line Case

Proof of Theorem 1 First of all, let us show that it would be sufficient to solve our problem in $L^2(\mathbb{R})$. Indeed, if $u(x)$ is a square integrable solution of (1.2), directly from this equation under our assumptions we obtain that

$$-\frac{d^2u}{dx^2} - b\frac{du}{dx} \in L^2(\mathbb{R})$$

as well. By using the standard Fourier transform (1.4), we arrive at $(p^2 - ibp)\widehat{u}(p) \in L^2(\mathbb{R})$. Hence, $\displaystyle\int_{-\infty}^{\infty} p^4|\widehat{u}(p)|^2 dp < \infty$, such that $\dfrac{d^2u}{dx^2} \in L^2(\mathbb{R})$. Therefore, $u(x) \in H^2(\mathbb{R})$ as well.

To show the uniqueness of solutions of (1.2), we suppose that $u_1(x)$, $u_2(x) \in L^2(\mathbb{R})$ satisfy this equation. Then their difference $w(x) := u_1(x) - u_2(x) \in L^2(\mathbb{R})$ is a solution of the homogeneous problem

$$\frac{d^2w}{dx^2} - b\frac{dw}{dx} - aw = 0.$$

Since the operator $L_{a,b}$, which is defined in (1.3) does not possess any nontrivial square integrable zero modes on the real line, the function $w(x)$ vanishes on $\mathbb{R}$.

We apply the standard Fourier transform to both sides of equation (1.2). This yields

$$\widehat{u}(p) = \frac{\widehat{f}(p)}{p^2 - a - ibp}. \tag{2.1}$$

Therefore,

$$\|u\|_{L^2(\mathbb{R})}^2 = \int_{-\infty}^{\infty} \frac{|\widehat{f}(p)|^2}{(p^2 - a)^2 + b^2 p^2} dp. \tag{2.2}$$

Let us first consider the case (a) of the theorem. By means of (2.2), we obtain

$$\|u\|_{L^2(\mathbb{R})}^2 \leq \frac{1}{C} \|f\|_{L^2(\mathbb{R})}^2 < \infty$$

due to the one of our assumptions. Here and throughout the article C will denote a finite, positive constant.

Then we turn our attention to the situation when $a = 0$. From (2.1), we easily express

$$\widehat{u}(p) = \frac{\widehat{f}(p)}{ib(p - ib)} - \frac{\widehat{f}(p)}{ibp}. \tag{2.3}$$

The first term in the right side of (2.3) is square integrable, since

$$\int_{-\infty}^{\infty} \frac{|\widehat{f}(p)|^2}{b^2(p^2 + b^2)} dp \leq \frac{1}{b^4} \|f\|_{L^2(\mathbb{R})}^2 < \infty$$

as assumed. The second term in the right side of (2.3) can be written as

$$\frac{i\widehat{f}(p)}{bp} \chi_{\{|p| \leq 1\}} + \frac{i\widehat{f}(p)}{bp} \chi_{\{|p| > 1\}}. \tag{2.4}$$

Here and further down χ_A will denote the characteristic function of a set $A \subseteq \mathbb{R}$. Clearly, the second term in (2.4) can be estimated from above in the absolute value by $\frac{|\widehat{f}(p)|}{|b|} \in L^2(\mathbb{R})$ since $f(x)$ is square integrable according to our assumption. Let us express

$$\widehat{f}(p) = \widehat{f}(0) + \int_0^p \frac{d\widehat{f}(s)}{ds} ds.$$

Hence, the first term in (2.4) can be written as

$$\frac{i\widehat{f}(0)}{bp} \chi_{\{|p| \leq 1\}} + \frac{i \int_0^p \frac{d\widehat{f}(s)}{ds} ds}{bp} \chi_{\{|p| \leq 1\}}. \tag{2.5}$$

Using definition (1.4), we easily estimate

$$\left| \frac{d\widehat{f}(p)}{dp} \right| \leq \frac{1}{\sqrt{2\pi}} \|xf(x)\|_{L^1(\mathbb{R})}.$$

Thus, the second term in (2.5) can be bounded from above in the absolute value by

$$\frac{1}{\sqrt{2\pi}|b|}\|xf(x)\|_{L^1(\mathbb{R})}\chi_{\{|p|\leq 1\}} \in L^2(\mathbb{R}).$$

Obviously, the first term in (2.5) is square integrable if and only if $\widehat{f}(0)$ vanishes, which is equivalent to orthogonality relation (1.8). ∎

Then we proceed to establishing the solvability in the sense of sequences for our problem on the real line.

Proof of Theorem 2 First of all, we suppose that Eqs. (1.2) and (1.5) admit unique solutions $u(x) \in H^2(\mathbb{R})$ and $u_m(x) \in H^2(\mathbb{R})$, $m \in \mathbb{N}$ respectively, such that $u_m(x) \to u(x)$ in $L^2(\mathbb{R})$ as $m \to \infty$. This will imply that $u_m(x)$ also converges to $u(x)$ in $H^2(\mathbb{R})$ as $m \to \infty$. Indeed, from (1.2) and (1.5) we easily obtain

$$\left\| -\frac{d^2(u_m - u)}{dx^2} - b\frac{d(u_m - u)}{dx} \right\|_{L^2(\mathbb{R})} \leq \|f_m - f\|_{L^2(\mathbb{R})} + a\|u_m - u\|_{L^2(\mathbb{R})}.$$

The right side of the inequality above tends to zero as $m \to \infty$ due to our assumptions. By applying the standard Fourier transform (1.4), we arrive at

$$\int_{-\infty}^{\infty} p^4 |\widehat{u}_m(p) - \widehat{u}(p)|^2 dp \to 0, \quad m \to \infty,$$

such that $\dfrac{d^2 u_m}{dx^2} \to \dfrac{d^2 u}{dx^2}$ in $L^2(\mathbb{R})$ as $m \to \infty$. Therefore, $u_m(x) \to u(x)$ in $H^2(\mathbb{R})$ as $m \to \infty$ as well.

By applying the standard Fourier transform (1.4) to both sides of (1.5), we arrive at

$$\widehat{u}_m(p) = \frac{\widehat{f}_m(p)}{p^2 - a - ibp}, \quad m \in \mathbb{N} \tag{2.6}$$

Let us first consider the case (a) of the theorem. By means of the part (a) of Theorem 1, Eqs. (1.2) and (1.5) admit unique solutions $u(x) \in H^2(\mathbb{R})$ and $u_m(x) \in H^2(\mathbb{R})$, $m \in \mathbb{N}$ respectively. By virtue of (2.6) along with (2.1), we derive

$$\|u_m - u\|_{L^2(\mathbb{R})}^2 = \int_{-\infty}^{\infty} \frac{|\widehat{f}_m(p) - \widehat{f}(p)|^2}{(p^2 - a)^2 + b^2 p^2} dp.$$

Hence,

$$\|u_m - u\|_{L^2(\mathbb{R})} \leq \frac{1}{C}\|f_m - f\|_{L^2(\mathbb{R})} \to 0, \quad m \to \infty,$$

which implies that in the case of $a > 0$ we have $u_m(x) \to u(x)$ in $H^2(\mathbb{R})$ as $m \to \infty$ as discussed above.

Let us conclude the proof of the theorem by considering the situation when the parameter $a = 0$. By means of the result of the part (a) of Lemma 3.3 of [21], under our assumptions we obtain that

$$(f(x), 1)_{L^2(\mathbb{R})} = 0 \tag{2.7}$$

holds. Then by virtue of the part (b) of Theorem 1, problems (1.2) and (1.5) have unique solutions $u(x) \in H^2(\mathbb{R})$ and $u_m(x) \in H^2(\mathbb{R})$, $m \in \mathbb{N}$ respectively when a vanishes. Formulas (2.6) and (2.1) give us

$$\widehat{u}_m(p) - \widehat{u}(p) = \frac{\widehat{f}_m(p) - \widehat{f}(p)}{ib(p - ib)} - \frac{\widehat{f}_m(p) - \widehat{f}(p)}{ibp},$$

which yields

$$\|u_m - u\|_{L^2(\mathbb{R})} \leq \frac{1}{|b|} \left\| \frac{\widehat{f}_m(p) - \widehat{f}(p)}{p - ib} \right\|_{L^2(\mathbb{R})} + \frac{1}{|b|} \left\| \frac{\widehat{f}_m(p) - \widehat{f}(p)}{p} \right\|_{L^2(\mathbb{R})}. \tag{2.8}$$

Clearly, the norm involved in the first term in the right side of (2.8) can be bounded from above by

$$\frac{1}{|b|} \|f_m - f\|_{L^2(\mathbb{R})} \to 0, \quad m \to \infty$$

as assumed. Let us express

$$\frac{\widehat{f}_m(p) - \widehat{f}(p)}{p} = \frac{\widehat{f}_m(p) - \widehat{f}(p)}{p} \chi_{\{|p| \leq 1\}} + \frac{\widehat{f}_m(p) - \widehat{f}(p)}{p} \chi_{\{|p| > 1\}}.$$

Hence

$$\left\| \frac{\widehat{f}_m(p) - \widehat{f}(p)}{p} \right\|_{L^2(\mathbb{R})} \leq$$

$$\leq \left\| \frac{\widehat{f}_m(p) - \widehat{f}(p)}{p} \chi_{\{|p| \leq 1\}} \right\|_{L^2(\mathbb{R})} + \left\| \frac{\widehat{f}_m(p) - \widehat{f}(p)}{p} \chi_{\{|p| > 1\}} \right\|_{L^2(\mathbb{R})}. \tag{2.9}$$

Evidently, the second term in the right side of (2.9) can be estimated from above by

$$\|f_m - f\|_{L^2(\mathbb{R})} \to 0, \quad m \to \infty$$

due to the one of our assumptions. Orthogonality conditions (2.7) and (1.9) give us

$$\widehat{f}(0) = 0, \quad \widehat{f}_m(0) = 0, \quad m \in \mathbb{N},$$

such that

$$\widehat{f}(p) = \int_0^p \frac{d\widehat{f}(s)}{ds} ds, \quad \widehat{f_m}(p) = \int_0^p \frac{d\widehat{f_m}(s)}{ds} ds, \quad m \in \mathbb{N}. \qquad (2.10)$$

Hence, it remains to estimate the norm of the term

$$\frac{\int_0^p [\frac{d\widehat{f_m}(s)}{ds} - \frac{d\widehat{f}(s)}{ds}] ds}{p} \chi_{\{|p| \leq 1\}}.$$

Using the definition of the standard Fourier transform (1.4), we easily derive

$$\left| \frac{d\widehat{f_m}(p)}{dp} - \frac{d\widehat{f}(p)}{dp} \right| \leq \frac{1}{\sqrt{2\pi}} \|xf_m(x) - xf(x)\|_{L^1(\mathbb{R})}.$$

Thus,

$$\left\| \frac{\widehat{f_m}(p) - \widehat{f}(p)}{p} \chi_{\{|p| \leq 1\}} \right\|_{L^2(\mathbb{R})} \leq \frac{1}{\sqrt{\pi}} \|xf_m(x) - xf(x)\|_{L^1(\mathbb{R})} \to 0, \quad m \to \infty$$

as assumed. Therefore, $u_m(x) \to u(x)$ in $L^2(\mathbb{R})$ as $m \to \infty$, which implies that $u_m(x) \to u(x)$ in $H^2(\mathbb{R})$ as $m \to \infty$ as discussed above. ∎

3   The Problem on the Finite Interval

Proof of Theorem 3 First of all, we prove that it would be sufficient to solve our problem in $L^2(I)$. Indeed, if $u(x)$ is a square integrable solution of (1.10), periodic on I along with its first derivative, directly from our equation under the given conditions we arrive at

$$-\frac{d^2u}{dx^2} - b\frac{du}{dx} \in L^2(I).$$

Using (1.12), we obtain $(n^2 - ibn)u_n \in l^2$. Thus, $\sum_{n=-\infty}^{\infty} n^4 |u_n|^2 < \infty$, such that $\frac{d^2u}{dx^2} \in L^2(I)$. This implies that $u(x) \in H^2(I)$ as well.

To prove the uniqueness of solutions of (1.10), we treat the case of $a > 0$. When a vanishes, we are able to use the similar argument in the constrained subspace $H_0^2(I)$. Let us suppose that $u_1(x)$, $u_2(x) \in H^2(I)$ solve (1.10). Then their difference $w(x) := u_1(x) - u_2(x) \in H^2(I)$ satisfies the homogeneous equation

$$\frac{d^2w}{dx^2} - b\frac{dw}{dx} - aw = 0.$$

Since the operator $\mathcal{L}_{a,\,b} : H^2(I) \to L^2(I)$ defined in (1.13) does not have any non-trivial $H^2(I)$ zero modes, the function $w(x)$ vanishes on I.

Let us apply the Fourier transform (1.12) to both sides of problem (1.10). This yields

$$u_n = \frac{f_n}{n^2 - a - ibn}, \quad n \in \mathbb{Z}. \tag{3.1}$$

First, we consider the case (a) of our theorem. By virtue of (3.1), we arrive at

$$\|u\|_{L^2(I)}^2 \le \frac{1}{C}\|f\|_{L^2(I)}^2 < \infty$$

via the one of our assumptions (see (1.11)). Let us conclude the proof of the theorem by treating the situation when $a = 0$. From (3.1), we easily derive

$$u_n = \frac{f_n}{ib(n - ib)} - \frac{f_n}{ibn}, \quad n \in \mathbb{Z}. \tag{3.2}$$

The first term in the right side of (3.2) belongs to l^2, since

$$\sum_{n=-\infty}^{\infty} \frac{|f_n|^2}{b^2(n^2 + b^2)} \le \frac{1}{b^4}\|f\|_{L^2(I)}^2 < \infty$$

as discussed above. The second term in the right side of (3.2) belongs to l^2 if and only if $f_0 = 0$ and the square of its l^2 norm can be easily bounded from above by $\frac{1}{b^4}\|f\|_{L^2(I)}^2 < \infty$, which is equivalent to orthogonality condition (1.16). ∎

Let us conclude the article with proving the solvability in the sense of sequences for our problem on the interval I with periodic boundary conditions.

Proof of Theorem 4 Evidently,

$$|f(0) - f(2\pi)| \le |f(0) - f_m(0)| + |f_m(2\pi) - f(2\pi)| \le 2\|f_m - f\|_{L^\infty(I)} \to 0$$

as $m \to \infty$. Hence, $f(0) = f(2\pi)$. By means of (1.11) under our assumptions, we have $f_m(x) \in L^1(I) \cap L^2(I), \quad m \in \mathbb{N}$. Using (1.11), we arrive at

$$\|f_m(x) - f(x)\|_{L^1(I)} \le 2\pi\|f_m(x) - f(x)\|_{L^\infty(I)} \to 0, \quad m \to \infty \tag{3.3}$$

as assumed, such that $f_m(x) \to f(x)$ in $L^1(I)$ as $m \to \infty$. Similarly, via (1.11) we obtain

$$\|f_m(x) - f(x)\|_{L^2(I)} \le \sqrt{2\pi}\|f_m(x) - f(x)\|_{L^\infty(I)} \to 0, \quad m \to \infty, \tag{3.4}$$

such that $f_m(x) \to f(x)$ in $L^2(I)$ as $m \to \infty$ as well. By applying the Fourier transform (1.12) to both sides of (1.15), we derive

$$u_{m,n} = \frac{f_{m,n}}{n^2 - a - ibn}, \quad m \in \mathbb{N}, \quad n \in \mathbb{Z}. \tag{3.5}$$

First we consider the case (a) of our theorem. By virtue of the part (a) of Theorem 3, problems (1.10) and (1.15) possess unique solutions $u(x) \in H^2(I)$ and $u_m(x) \in H^2(I)$, $m \in \mathbb{N}$ respectively. (3.5) along with (3.1) yield

$$\|u_m - u\|^2_{L^2(I)} = \sum_{n=-\infty}^{\infty} \frac{|f_{m,n} - f_n|^2}{(n^2 - a)^2 + b^2 n^2} \leq \frac{1}{C} \|f_m - f\|^2_{L^2(I)} \to 0, \quad m \to \infty$$

via (3.4). Thus, $u_m(x) \to u(x)$ in $L^2(I)$ as $m \to \infty$. Let us show that $u_m(x)$ converges to $u(x)$ in $H^2(I)$ as $m \to \infty$. Indeed, (1.10) and (1.15) give us

$$\left\| -\frac{d^2(u_m - u)}{dx^2} - b\frac{d(u_m - u)}{dx} \right\|_{L^2(I)} \leq \|f_m - f\|_{L^2(I)} + a\|u_m - u\|_{L^2(I)}.$$

The right side of this inequality converges to zero as $m \to \infty$ due to (3.4). Using the Fourier transform (1.12), we derive

$$\sum_{n=-\infty}^{\infty} n^4 |u_{m,n} - u_n|^2 \to 0, \quad m \to \infty.$$

Hence, $\dfrac{d^2 u_m}{dx^2} \to \dfrac{d^2 u}{dx^2}$ in $L^2(I)$ as $m \to \infty$, such that $u_m(x) \to u(x)$ in $H^2(I)$ as $m \to \infty$ as well.

We conclude the article by dealing with the situation when the parameter a vanishes. By means of (1.17) along with (3.3), we obtain

$$|(f(x), 1)_{L^2(I)}| = |(f(x) - f_m(x), 1)_{L^2(I)}| \leq \|f_m - f\|_{L^1(I)} \to 0, \quad m \to \infty,$$

such that

$$(f(x), 1)_{L^2(I)} = 0 \tag{3.6}$$

holds. By virtue of the part b) of Theorem 3 above Eqs. (1.10) and (1.15) admit unique solutions $u(x) \in H^2_0(I)$ and $u_m(x) \in H^2_0(I)$, $m \in \mathbb{N}$ respectively when $a = 0$. Formulas (3.1) and (3.5) yield

$$u_{m,n} - u_n = \frac{f_{m,n} - f_n}{ib(n - ib)} - \frac{f_{m,n} - f_n}{ibn}, \quad m \in \mathbb{N}, \quad n \in \mathbb{Z}. \tag{3.7}$$

By means of orthogonality relations (3.6) and (1.17), we have

$$f_0 = 0, \quad f_{m,0} = 0, \quad m \in \mathbb{N}.$$

Obviously, the l^2 norm of the first term in the right side of (3.7) can be bounded from above by

$$\frac{1}{b^2} \| f_m - f \|_{L^2(I)} \to 0, \quad m \to \infty$$

via (3.4). The l^2 norm of the second term in the right side of (3.7) can be estimated from above by

$$\sqrt{\sum_{n=-\infty,\, n\neq 0}^{\infty} \frac{|f_{m,n} - f_n|^2}{b^2 n^2}} \leq \frac{1}{|b|} \| f_m - f \|_{L^2(I)} \to 0, \quad m \to \infty$$

as above. Therefore, $u_m(x) \to u(x)$ in $L^2(I)$ as $m \to \infty$, which implies that $u_m(x) \to u(x)$ in $H_0^2(I)$ as $m \to \infty$ as well via the argument analogous to the one we had in the proof of the part (a) of the theorem.   ∎

References

1. Amrouche C, Girault V, Giroire J (1997) Dirichlet and Neumann exterior problems for the n-dimensional Laplace operator: an approach in weighted Sobolev spaces. J Math Pures Appl 76(1):55–81
2. Amrouche C, Bonzom F (2008) Mixed exterior Laplace's problem. J Math Anal Appl 338(1):124–140
3. Benkirane N (1988) Propriétés d'indice en théorie hölderienne pour des opérateurs elliptiques dans R^n. C R Acad Sci Paris Sér I Math 307(11):577–580
4. Berestycki H, Hamel F, Nadirashvili N (2005) The speed of propagation for KPP type problems I Periodic framework. J Eur Math Soc (JEMS) 7(2):173–213
5. Bolley P, Pham TL (1993) Propriétés d'indice en théorie hölderienne pour des opérateurs différentiels elliptiques dans R^n. J Math Pures Appl 72(1):105–119
6. Bolley P, Pham TL (2001) Propriété d'indice en théorie Hölderienne pour le problème extérieur de Dirichlet. Comm Partial Differ Equ 26(1–2):315–334
7. Ducrot A, Marion M, Volpert V (2005) Systemes de réaction-diffusion sans propriété de Fredholm. C R Math Acad Sci Paris 340(9):659–664
8. Ducrot A, Marion M, Volpert V (2008) Reaction-diffusion problems with non-Fredholm operators. Adv Differ Equ 13(11–12):1151–1192
9. Hislop PD, Sigal IM (1996) Introduction to spectral theory with applications to Schrödinger operators. Applied Mathematical Sciences, 113. Springer, Verlag, p 337
10. Volpert V (2011) Elliptic partial differential equations, volume 1 Fredholm theory of elliptic problems in unbounded domains. Monographs in Mathematics, 101. Birkhäuser, Springer, 639 pp
11. Volpert V, Kazmierczak B, Massot M, Peradzynski Z (2002) Solvability conditions for elliptic problems with non-Fredholm operators. Appl Math (Warsaw) 29(2):219–238
12. Vougalter V, Volpert V (2010a) Solvability relations for some non Fredholm operators. Int Electron J Pure Appl Math 2(1):75–83

13. Vougalter V, Volpert V (2010b) On the solvability conditions for some non Fredholm operators. Int J Pure Appl Math 60(2):169–191
14. Vougalter V, Volpert V (2011a) Solvability conditions for some non-Fredholm operators. Proc Edinb Math Soc 54(1):249–271
15. Vougalter V, Volpert V (2011b) On the existence of stationary solutions for some non-Fredholm integro-differential equations. Doc Math 16:561–580
16. Volpert V, Vougalter V (2011c) On the solvability conditions for a linearized Cahn-Hilliard equation. Rend Istit Mat Univ Trieste 43:1–9
17. Vougalter V, Volpert V (2012a) On the solvability conditions for the diffusion equation with convection terms. Commun Pure Appl Anal 11(1):365–373
18. Vougalter V, Volpert V (2012b) Solvability conditions for a linearized Cahn-Hilliard equation of sixth order. Math Model Nat Phenom 7(2):146–154
19. Vougalter V, Volpert V (2012c) Solvability conditions for some linear and nonlinear non-Fredholm elliptic problems. Anal Math Phys 2(4):473–496
20. Volpert V, Vougalter V (2013a) Emergence and propagation of patterns in nonlocal reaction-diffusion equations arising in the theory of speciation. Dispersal, individual movement and spatial ecology. Lecture Notes in Math 2071. Springer, Heidelberg, pp 331–353
21. Volpert V, Vougalter V (2013b) Solvability in the sense of sequences to some non-Fredholm operators. Electron J Differ Equ 160:16
22. Vougalter V, Volpert V (2018) Existence in the sense of sequences of stationary solutions for some non-Fredholm integro-differential equations. J Math Sci (NY) 228(6):601–632. Problems in mathematical analysis, vol 90 (Russian)

NONLINEAR PHYSICAL SCIENCE

(Series Editors: Albert C.J. Luo, Dimitri Volchenkov)

ISBN 978-7-04-050235-0	42 Rigid Body Dynamics: Hamiltonian Methods, Integrability, Chaos (2018) by A. V. Borisov, I. S. Mamaev
ISBN 978-7-04-048458-8	41 Galloping Instability to Chaos of Cables (2018) by Albert C. J. Luo, Bo Yu
ISBN 978-7-04-048004-7	40 Resonance and Bifurcation to Chaos in Pendulum (2017) by Albert C. J. Luo
ISBN 978-7-04-047940-9	39 Grammar of Complexity: From Mathematics to a Sustainable World (2017) by Dimitri Volchenkov
ISBN 978-7-04-047809-9	38 Type-2 Fuzzy Logic: Uncertain Systems' Modeling and Control (2017) by Rómulo Martins Antão, Alexandre Mota, R. Escadas Martins, J. Tenreiro Machado
ISBN 978-7-04-047450-3	37 Bifurcation in Autonomous and Nonautonomous Differential Equations with Discontinuities (2017) by Marat Akhmet, Ardak Kashkynbayev
ISBN 978-7-04-043231-2	36 离散和切换动力系统（中文版）(2015) 罗朝俊
ISBN 978-7-04-043102-5	35 Replication of Chaos in Neural Networks, Economics and Physics (2015) by Marat Akhmet, Mehmet Onur Fen
ISBN 978-7-04-042835-3	34 Discretization and Implicit Mapping Dynamics (2015) by Albert C.J.Luo
ISBN 978-7-04-042385-3	33 Tensors and Riemannian Geometry with Applications to Differential Equations (2015) by Nail Ibragimov
ISBN 978-7-04-042131-6	32 Introduction to Nonlinear Oscillations (2015) by Vladimir I. Nekorkin
ISBN 978-7-04-038891-6	31 Keller-Box Method and Its Application (2014) by K. Vajravelu, K.V. Prasad
ISBN 978-7-04-039179-4	30 Chaotic Signal Processing (2014) by Henry Leung
ISBN 978-7-04-037357-8	29 Advances in Analysis and Control of Time-Delayed Dynamical Systems (2013) by Jianqiao Sun, Qian Ding (Editors)
ISBN 978-7-04-036944-1	28 Lectures on the Theory of Group Properties of Differential Equations (2013) by L.V. Ovsyannikov (Author), Nail Ibragimov (Editor)

ISBN 978-7-04-031533-2	12 Bifurcation and Chaos in Discontinuous and Continuous Systems (2011) by Michal Fečkan
ISBN 978-7-04-031695-7	11 Waves and Structures in Nonlinear Nondispersive Media: General Theory and Applications to Nonlinear Acoustics (2011) by S.N. Gurbatov, O.V. Rudenko, A.I. Saichev
ISBN 978-7-04-032187-6	10动态域上的不连续动力系统 （中文版）(2011) 罗朝俊著，闵富红，黄健哲，郭羽译
ISBN 978-7-04-029474-3	9 Self-organization and Pattern-formation in Neuronal Systems Under Conditions of Variable Gravity (2011) by Meike Wiedemann, Florian P.M. Kohn, Harald Rosner, Wolfgang R.L. Hanke
ISBN 978-7-04-029473-6	8 Fractional Dynamics (2010) by Vasily E. Tarasov
ISBN 978-7-04-029187-2	7 Hamiltonian Chaos beyond the KAM Theory (2010) by Albert C.J. Luo, Valentin Afraimovich (Editors)
ISBN 978-7-04-029188-9	6 Long-range Interactions, Stochasticity and Fractional Dynamics (2010) by Albert C.J. Luo, Valentin Afraimovich (Editors)
ISBN 978-7-04-028882-7	5 Nonlinear Deformable-body Dynamics (2010) by Albert C.J. Luo
ISBN 978-7-04-018292-7	4 Mathematical Theory of Dispersion-Managed Optical Solitons (2010) by A. Biswas, D. Milovic, E. Matthew
ISBN 978-7-04-025480-8	3 Partial Differential Equations and Solitary Waves Theory (2009) by Abdul-Majid Wazwaz
ISBN 978-7-04-025759-5	2 Discontinuous Dynamical Systems on Time-varying Domains (2009) by Albert C.J. Luo
ISBN 978-7-04-025159-3	1 Approximate and Renormgroup Symmetries (2009) by Nail H. Ibargimov

郑重声明

高等教育出版社依法对本书享有专有出版权。任何未经许可的复制、销售行为均违反《中华人民共和国著作权法》，其行为人将承担相应的民事责任和行政责任；构成犯罪的，将被依法追究刑事责任。为了维护市场秩序，保护读者的合法权益，避免读者误用盗版书造成不良后果，我社将配合行政执法部门和司法机关对违法犯罪的单位和个人进行严厉打击。社会各界人士如发现上述侵权行为，希望及时举报，本社将奖励举报有功人员。

反盗版举报电话　（010）58581999　58582371
反盗版举报邮箱　dd@hep.com.cn
通信地址　北京市西城区德外大街4号
　　　　　高等教育出版社法律事务部
邮政编码　100120